T0219976

Raw Data

Pernille Rørth

Raw Data

A Novel on Life in Science

 Springer

Pernille Rørth
Copenhagen, Denmark

ISBN 978-3-319-23972-9   ISBN 978-3-319-23974-3   (eBook)
DOI 10.1007/978-3-319-23974-3

Library of Congress Control Number: 2015954181

Springer Cham Heidelberg New York Dordrecht London
© Springer International Publishing Switzerland 2016

Cover: Photo on front side by Danny E. E. Fievez. Figure credit for picture of the author on back cover by Danny E. E. Fievez.

Printed on acid-free paper

Springer International Publishing AG Switzerland is part of Springer Science+Business Media (www.springer.com)

Preface

I would like to thank the many scientists who have contributed to making my years in science so very interesting. I would also like to thank Christian Caron, my editor, for his excellent suggestions, and Steve, for everything.

This is a work of fiction.

Copenhagen, Denmark   Pernille Rørth
June 2015

Contents

Part I

The Novel

Raw Data

Chapter 1

December 2014

What she really needs is a coffee. Not the lunch bag that somehow ended up in her hand. No, something to keep her going now that the talk-fueled adrenaline is starting to wane. Standing in line, Karen glances around the large convention hall. The poster session is still busy after lunch, but seems less intense than the previous days. People may be getting tired, she thinks, information overload setting in after four packed days. She should try for some posters, though. Or at least mingle. Bad conference coffee in hand, she heads into the crowd.

She scans the poster titles as she wanders down the first row. A few dramatic pictures and colorful diagrams catch her eye. But nothing holds her attention for long. Perhaps meeting fatigue is finally getting to her as well. Turning the corner, she finds a small crowd filling the aisle. A popular poster, it seems. The presenter is an energetic woman in her late twenties, of medium height and build and with mousy blond hair. Karen smiles, reminded of herself attending her first international meeting, years and years ago. She steps closer, trying to hear above the background noise. As she does, a man turns around.

"Karen Larsson. Ah, finally. It's great to see you." He holds out his hand to her and continues rapidly. "I really enjoyed your talk this morning. Wonderful stuff. Really amazing." He is about her age, half a head taller, with curly red-blond hair and beard. He seems familiar, but no name springs to mind immediately. Karen smiles and shakes his hand with a tentative "Thanks". After a moment she realizes that he spoke with the familiar sing-song accent of home. Now she remembers. Torsten. He did his PhD in Nils' lab at Karolinska, starting a year after her.

"Torsten. What a surprise. I haven't seen you for like, forever. How are you?" They fall into an easy chat, catching up on the missing 10 years. He is

© Springer International Publishing Switzerland 2016
P. Rørth, *Raw Data*, DOI 10.1007/978-3-319-23974-3_1

still in Sweden, in Lund. Her lab is in Chicago. Prompted, she tells him about her postdoc in Boston, the job at Northwestern, her PhD students and their projects. She has come far, she knows, and feels the rush of well-earned pride. He compliments her on the talk again and the "fantastic movies" she showed. This allows her to gush happily about her "favorite toy", the high-end, custom-equipped microscope that she has spent most of the past 5 years with. After a while, they run out of obvious topics. But saying goodbye, she finds herself re-energized and ready for more.

Determined to make an effort for the rest of the session, she moves on, looking closely at a few posters, listening to bits of presentations here and there. She manages to pick up a few interesting findings, but mostly drifts along. Several times, she receives a "Great talk this morning" or "I really enjoyed your talk" from complete strangers. This is immensely gratifying. It was worth staying on this afternoon, she thinks, just for the pleasure of these spontaneous compliments.

Halfway down the next aisle, she stops up, abruptly. The perfect hair, the confident walk. Chloe. It must be her. Karen freezes, unable to move forward or to turn away. Then the woman turns to face her. It is not Chloe. Of course not. Pull yourself together, she thinks. You thought she was at the talk too, in the third row. But that wasn't her either. She is not here. Karen breathes deeply. It is her first time back since she left Boston some 5 years ago. That must be what is making her jumpy. Silly.

Slowly, she starts moving again. She reads more poster titles and glances at images and text. She gets another compliment on her talk. After some time, she notices that the hall is slowly starting to empty out. The first posters are being taken down, rolled up and returned to their protective plastic tubes. A few eager students are still explaining their work. Others are simply chatting, reluctant to let it all go, perhaps. She remembers her lunch bag and eats half of the slightly stale sandwich. It's hard to believe it has already come and gone, her first talk at a major meeting, something she had been looking forward to for so long. It was worth it, though, all the years, all the preparations. She thinks back to being on the podium, being able to engage that big audience, and the questions and compliments after. Yes, definitely worth it. But the moment has passed and the meeting is closing. She takes a final look around, scanning the faces close by. There are no ghosts here, only ordinary strangers and the occasional old friend. It is better this way. The past is safe where it is.

Chapter 2

November, 2006

The polished red stone is shimmering like a faint halo in the pale morning sunlight. Chloe loves this building. Although smaller than the gray hulks surrounding it, it immediately attracts the eye: the unexpected color and light, the subtle, yet confident curves. The adjacent university buildings seem bland, uninspired by comparison. Chloe feels that she truly belongs here now, at the institute. She has earned her place, even if she is only a member of the transient postdoc population. Four years ago she was a newbie full of well-suppressed insecurities, in awe of the famous faculty members. That girl seems so remote now. Since then, the institute has become her home, her world. As intense and challenging and full a world as she had hoped it would be.

She takes a few quick steps to get to the entrance. However much she loves it here, she is ready for the next phase. It is time to move on and start her own lab as a principal investigator. Preferably at an equally excellent research institute or university. She hopes that today is the day for a final decision about her paper. It has been 3 weeks since the resubmission to Nature. The initial reviews were basically positive and she has done everything the reviewers asked for. So the journal editors must accept her paper, they must. The truth is that they can do what they want, and she knows it. But this morning she feels convinced that it will be a yes. She quickens her step and pushes open the huge glass doors with her shoulder, shielding the still warm latte in one hand.

On her way in, Chloe smiles and says "Good morning" to Mr. Cleveland, one of the usual security guys. He is a large man and his fitted light blue uniform makes him look even more imposing. But he is always ready with a friendly smile and is quick to buzz Chloe in when he is on weekend shifts. On this Friday morning, the door is not locked, but a nod and a smile are offered as she passes through the glassed-in entryway. Chloe appreciates the easy friendliness of most Americans. It is so much better than the routine German grumpiness that she grew up with. Nosy watchers always looking for impropriety in a smile or an attractive face. Although not conventionally pretty, Chloe is certainly attractive. She has a handsome, angular face with prominent cheekbones, framed by short, expertly cut, dark brown hair. She is slender, well proportioned and athletic. Partial to ankle boots, blue jeans and fitted cotton shirts, she dresses casually but never sloppily. She is content that she has found a good balance. Feminine enough, yet she does not draw unwanted attention. Her freshness, confidence and intelligent charm are what tend to draw people in, and what most remember. This suits her.

Once inside, Chloe fleetingly admires the building's sleek interior design and gives the art by the elevators the usual quick glance. But she does not linger. She heads for the stairs. The elevators are maddeningly slow, as if she has all the time in the world to get to the lab and get going.

The third floor corridor has glass all along one side, offering a full view of the labs. She initially found this fish tank exposure unnerving. But almost no one wastes time standing in these corridors, looking in. Passing by, she glances into the Wu lab. It is busy, as usual. Heads of black, straight hair are all bent in concentration. The next room along is her lab. Well, it's Tom Palmer's lab. But it has been her home these past few years. It looks just like the Wu lab, the same benches, the same desks, even the same centrifuges, gel-boxes and row upon row of clear bottles on the shelves above. But Tom's lab has more night owls like her, so not everyone is in yet.

Juan is in, working already. He occupies the bench next to Chloe's, so she slides by him to get to her desk. She smiles with a friendly "Good morning, Juan" and he counters with his standard "Ola". A family man, he is always in before her and usually leaves around five in the afternoon. This suits Chloe. She then has both benches at her disposal in the evening, should she need them for experiments. Also, because of his time-constraints, Juan rarely fools around in the lab. An optimal lab neighbor.

This morning, she notices Juan's gym bag underneath his desk. Odd. His regular handball practice is on Wednesdays, not Fridays. She plays with the women's team on Thursdays. When she moved here, she was pleasantly surprised to find that Boston hosted teams for this mostly European ball game. She never got around to trying basketball, even if it does look like fun. Handball is a bit rougher, but she is fast, strong and unafraid, crucial attributes for the game. So the local team is happy to have her. For both Chloe and Juan, matches and tournaments take place on weekends. They sometimes run into each other at these events. They watch the games and admire particularly good jump-shots (Chloe's specialty) or impossibly twisted shots from the line (more Juan's style). Sometimes they talk about the lab as well. The first time they saw each other at the indoor courts, they were both taken aback. Somehow one doesn't expect to see lab mates outside of the lab.

"Is there a match tonight?" She asks, nodding toward the bag.

"No, it's for the weekend. I was planning to go play a bit during my incubations tomorrow."

"So the home front will think you are hard at work the whole time? Very tricky, Juan. I didn't think you had it in you."

"It's a two-hour incubation. I can't make it home and back in that time." He peers at Chloe and realizes she is teasing. "Do you want to come? We need a few more players."

"I might do. I should be able to handle you wimpy boys. I'll bring in my stuff tomorrow. Let me know when you head off?"

"Sure." Juan flashes a quick smile and turns back to his work.

Chloe's bench is unusually orderly. Normally it would feature abandoned pipetmen, boxes of blue and yellow tips and scattered colorful racks of recently used Eppendorf tubes. But today, nothing is out of place. Her desk is also neat and tidy. The tidiness is a result of the inevitable displacement activity before a major writing job. For the past couple of weeks she has forced herself to stay at her computer and finalize her research proposal. So her fingers are now itching to get going on experiments again. But the writing time has been productive and she knows the proposal is the most important part of her job application package. Well, it will be, once her paper is in press. Maybe today. She looks down to the end of the long room, past the many lab benches, and notices that the door to Tom's office is half-open. This means that he is in. He may have gotten word from Nature and sent it on to her.

She sends a quick wish to the nonexistent Gods of hardworking postdocs. User-name, password—and a deep breath before hitting the return button. There it is, forwarded from Tom half an hour ago: *'Decision on your manuscript N-06-22881'*. One email, so simple. The final judgment on her ideas and 4 years of fiercely hard work. Her heart beats faster. She sits down before double-clicking. This is it.

'Dear Dr. Palmer, we are pleased to inform you that your manuscript N-06-22881 has been accepted for publication . . .'. "Yes, yes" she blurts out. Juan turns around.

"Juan, my paper just got accepted." She says immediately, smiling, smiling; she is ecstatic. ". . . at Nature, as a full article".

"That's fantastic. Congratulations." He seems truly pleased for her.

"It is, it is". Chloe leans back in her chair, letting it sink in. Happiness spreads through her, followed by relief. What she has been hoping for—what she needs—is right here. Her paper accepted for publication. The ultimate success of her ambitious project, built on her risky ideas. She never imagined so much work could go into a single paper. Four years, working like crazy. So much trial and error, so much uncertainty. Each little step forward a minor victory, necessary fuel to keep going. Several times along the way she felt that she had enough for a good paper. But Tom nudged her on. "Chloe, what you have is good, yes, but this one could be great. You have to hit it out of the park." So the satisfaction now is immense—as is the relief. It is a home run. And it is all hers. She knows that the work, as published, will look completely logical and straightforward, each of the many steps is a simple progression from the one before. There is no sign of the many muddles and delays along the way. Her paper describes a well-designed, successful journey, impressive in effort and important in findings. This is how she will tell the story in her job talks.

She reads the rest of the forwarded email. It is a form letter, more or less. She has seen such letters before. The only word that matters is "accepted". A '*Dear Dr. Varga,..*' would be even better, of course. But that will come. Tom is always corresponding author on papers from his lab. Admittedly, he did come with some good suggestions along the way. Once Myc was involved, he became seriously interested in the project. When she is on her own, she will not have to share the credit. Later. For now, she should find Tom.

On the way to Tom's office, Chloe passes Michel at his bench. His curly head is bent over a gel he is loading, apparently deeply engrossed in this simple task. She moves closer, her exuberance spilling over.

"Guess what, Michel? My paper got into Nature".

He lifts his head, his still loaded pipetman in one blue-gloved hand, and the now empty tube in the other.

"Félicitations. Fantastic, Chloe. C'est vraiment merveilleux. The Myc demethylase paper?" Michel's thick accent transforms the potent oncogene to 'meek".

"Yes, Oui, naturellement. I am so happy." She beams.

Michel did his PhD in France, so coming to Tom's lab for a postdoc was a major change. He seems happy now, but it was bumpy in the beginning, Chloe remembers. When he found that she spoke a reasonable French and had read some of his favorite authors without the benefit of translation, he opened up. With the many hours in the lab, they occasionally have time for a coffee and a chat. For a short while, Michel leaves his struggles with English for the pleasure of his mother tongue and Chloe battles with yet another language from her ambitious, if unfocused, youth. A simple friendship of shared words.

"Right. I'd better go see Tom." She leaves Michel to his gel-loading.

From Tom's office she hears the usual semi-furious tapping. She knocks on the open door and steps in. A full smile breaks through his expression of concentrated intent when he sees her. Tom is in his mid-fifties, fit and always has a bit of a tan. What is left of his hair is cropped short. There must be some gray in there, but it is hard to tell. He almost always wears blue jeans and a faded T-shirt. Given his position, this initially seemed to Chloe to be too informal. But everyone, including the occasional expensively suited trustee, seems to find this attire charming. The overly casual clothes are probably deliberate, Tom's little game. Certainly, it would be a mistake to take the relaxed look as an indication that he is laid back. He has an intensity that helps him steer his lab full of intelligent individualists and maintain his position among sharp-witted peers in the bigger world. It is a manner that Chloe has been watching and has learned from. Today, Tom appears to be truly content, for once.

"So, you saw the email?"

"Yes, I did. I'm so happy, ecstatic. We made it."

"Well, they would have been total idiots not to accept it at this point. It is a great paper and you deserve the credit. So congratulations, Chloe". Tom counters, generously. It is satisfying to hear Tom acknowledging this—her ideas, her work.

As Chloe sits in the chair opposite, Tom's phone rings.

"Hello?. . . Oh yes, Hi. Good of you to get back to me." Chloe starts to get up but Tom signals her to stay. To reinforce the intent, he covers the mouthpiece briefly and says. "This will only take a second." Tom turns halfway around to continue talking. Meanwhile Chloe lets her gaze wander the office, to be polite.

The first time she was in this office, ages ago, for her postdoc interview, she noticed nothing. She talked and talked, she was way too nervous. Tom nodded along, asking a question here and there. She had no idea what he was thinking or how she was doing. Only later in the day when he gave her a list of fellowships to apply for, did she realize that the interview must have gone well. Since that first stressful conversation, visits to the office have been brief and focused on the work at hand. So she has not paid much attention to the décor. There are almost no personal touches to Tom's office. There is a simple round table with two matching chairs, one occupied by a pile of journals. A giant bookcase takes up all of one wall. Chloe thinks she will do something bolder with her own office, once she gets one. She would start by hanging her two Clemente watercolors. Then add some more art worth looking at when she can afford it. So the place would really be hers. But maybe this neutral look is more professional. Something to consider.

Tom has turned around at his desk. With raised eyebrows and hand gestures he signals impatience with his phone conversation. It is OK for Chloe. For once she is not in a hurry. She studies the bookcase. Most of the space is taken up by row upon row of old lab notebooks, plus few bound PhD theses. The history of Tom's lab. All the data generated by former students and postdocs over the years, successful experiments as well as failed ones. The sheer mass of it draws her in. Tom also has another, more public version of lab accomplishments. These are cover images from lab publications that made it to the top journals Cell, Science or Nature. The big stories. These framed journal covers hang in the corridor. Showy, maybe too showy, she has thought in the past. But today it reminds her that she should think about a suggestion for cover image to submit to Nature. She should design something more original than the usual three-color antibody-stained cells. Something that cleverly illustrates the mechanism she uncovered. . .

Chloe is called back from her musing by Tom finally wrapping up his phone conversation.

"Sorry about that. Some people just go on and on. Anyway, nothing further that needs to be done with the paper, is there? I'll get the forms signed and sent off. It is a beautiful story. You should be very proud." He repeats.

"Thanks. I. . ."

"The institute will want to do some PR on this. You should get some simplified figures ready. And a suggestion for cover image, if you want."

"Yes, of course." These little extra jobs will be pure joy. She smiles again. Despite the massive significance of the paper and its acceptance to her, she realizes that there is little more to say about it now.

"I wanted to tell you that I am sending off the first job applications next week. They will be asking you for letters of recommendation."

She expects the all-important letter from Tom, her postdoc mentor, will be suitably enthusiastic. He is an American, after all, so the superlatives should flow.

"Excellent. Perfect timing. Where are you applying? You should aim high. You have what it takes and with this paper in press, you are in the sweet spot this year. So there is no reason to apply to second-tier places."

She tells him of the nine departments she is applying to. He smiles, nods, appreciating the ambition reflected in these choices.

"Perfect. You can always cast a broader net later on. But I doubt it will be necessary. You'll wow them, I'm sure. The interview process should also be an interesting experience for you. Some pretty smart folks will be paying close attention to your work and your ideas. Lots of excellent scientists at these places."

She has not dared to think about that part of the job-hunting process yet. Beyond the all-important job talk, interviews involve talking one-on-one to key scientists, about her work and about their work. At these places, that should be very exciting. But first, she needs to get invited.

Something else suddenly occurs to Chloe.

"What about the competing paper, from the White group? Is it coming out soon as well? You told me they were onto something similar."

"I don't think so. I don't think they were as far along."

Chloe is puzzled. She thought there was very close competition. Wasn't that what Tom said a few months ago? Well, this is not important now. "So it's all my story, I mean, all ours?"

"Yep. That it is. The patent application for L-334, and related inhibitors of Jmjd10 has also been filed. It should sail right through."

Chloe had almost forgotten about the patent application. It was a pain to go over what the patent lawyers gave them, so much repetition and detail. Maybe it will make money one day. Who knows? She cannot make herself care about that. No, she needs to get back to those job applications.

"We should celebrate properly" Tom says as Chloe gets up to leave. "Get some champagne and cake for the lab this afternoon. Let me give you some money."

He pulls out his wallet and counts out some 20's. Handing them over, he reminds her "First author buys, last author pays, right?"

So she should go shopping. Another chore she is quite content to do.

"Sure, I'll take care of that now. I'd better get the bubbles cooling for later. And, thanks for, well, thanks for everything."

On her way back to her desk to get her coat, she realizes she should tell Martin. Surely this is worth disturbing his morning for. Martin's lab, one of the small labs of the Institute Fellows, is on the fifth floor. Still full of energy, Chloe runs up the four half-flights of stairs and hurries down the corridor. She spots him, through the glass. He is standing behind his postdoc, Chen something, both of them looking at a computer screen. Chloe knocks gently on the glass and cocks her head with a smile as they look up. Martin seems not altogether pleased at the interruption. She enters the lab, deciding to ignore the hint of annoyance, but also to keep it short.

"I just wanted to say—my paper got accepted."

"Fantastic. Congratulations." Martin's expression changes. He smiles broadly. As expected, no kiss or hug is on offer—not at work. "I told you it would go straight in" he continues "didn't I?"

"You did, and you were right. Good for me."

"We should celebrate tonight. Call me later, OK?"

"Sure, of course. We will do something nice." Another charming smile. He does look happy for her. But he also wants to get back to work. He glances back down at Chen, who is waiting for a sign to continue.

"OK, see you later." she says, cheerfully, as she turns to leave.

Chloe remembers when Martin first came to the institute to give a seminar, applying to be an institute fellow. This was not your typical recent PhD seminar. It was almost irritating how in control he was. His smooth, yet thorough responses to the probing questions from the Institute's senior faculty were impressive. She managed to come up with a clever, unexpected question as well. So he noticed her and saw that she was to be taken seriously. It also broke the ice, giving her an excuse to talk to him afterwards. There might have been a girlfriend in the background, somewhere. But once he moved to

Boston, it did not take Chloe long to charm him. Or, perhaps, he charmed her. Whatever. The attraction and everything else was mutual. Apart from their dedication to their work, they found that they share cultural interests. Now, they laugh at themselves as a cliché of the perfect science couple.

Back at her desk, Chloe opens the folder with job application files. The satisfaction of replacing "submitted" with "Nature, in press" next to the title of her manuscript is enormous. She looks at her CV displayed on the screen. Not bad, she thinks. Four first author publications in good journals. The Nature paper is doubly important. It shows she can do top level science and the timing is perfect. Her work is now by definition 'important'. The proposal gets another read-though but nothing needs changing. She has worked hard on this, laboring over both the ideas and the presentation. But it is good now. It presents the importance of this specific sub-area of basic cancer research, her contributions so far and her plans and ideas for the future. And it does all this in just three pages, straightforward and logical. She decides to send the first applications off today, rather than waiting until the deadline. It is, after all, a perfect job for today.

The afternoon celebration in the lab goes off much as Chloe had expected. A lot of smiling and congratulations, some standing around and small talk. Her favorite brownies serve as "cake"—it is her party after all. And champagne. It had to be two bottles given a lab of almost 20 people. And it had to be real champagne, she decided, to set the right tone. Tom says a few words of praise, toasts Chloe and, as always, ends with a jolly "…and may you all be next in line." Meant to be encouraging, of course, but it has an undertone. The truth is, no one needs reminding.

Vikram soon starts fooling around, building architectural brownie constructions. Vik's combination of smarts and silliness is occasionally bewildering but never boring. Chloe welcomes this infusion of lightheartedness. The congratulations from him were also warm and genuine. From others, the same words were less convincing. She understands. She likes most of her lab mates, and she believes it is reciprocal. But, of course, they all want to be in her shoes. So there is some awkwardness as well, a bit of envy, and no use denying it.

Stealing a chimney from Vik's construction, Michel comes over to where Chloe is momentarily alone.

"Encore de félicitation. C'est merveilleux. Tu le mérites."

"Merci, Michel. Très gentil."

"Mais n'oublie pas tes amis quand tu deviens célèbre."

"T'oublier? Jamais."

They continue the light banter for a while. When Juan and Vik move closer, they switch into English but maintain the playful flirtatiousness that comes natural to both of them. It is, Chloe knows, quite innocent. She sips more champagne and enjoys their quartet's conversation, as it bounces around. The banter and laughter shared by this small group washes away the slightly strained feeling from earlier. Tom has already retreated to his office. From the rest of the room, people soon start drifting off, back to the lab. It is, after all, just a normal workday.

Chloe is still awake, savoring the last moments of the day. They are at her place. Martin is lying next to her, already asleep. It is hard to recognize the public Martin, the brilliant Martin, in this sleeping figure. She thinks of their dinner at Tosca, their private celebration. It was fun and pleasant, mostly. The future came up, naturally, since she talked about her job applications. It is always awkward. They can't help but get a touch defensive, both of them. But the situation is clear. She wants a good job, so she applies to the best places, wherever they are. He would do the same, in her shoes. She is sure of this. Just as she is sure that he will stay in his privileged position as an independent fellow until he is ready to move on to another great job. They respect each other too much to expect anything else. So usually they try to retain a light tone in any conversations about the future. He playfully says he will follow her wherever she goes, when the time is right. She pretends to believe him. It is an act they have and it normally relieves the tension. Tonight, though, her future has become just that bit more real. So it was harder to keep up the pretense. Maybe she should be sad. But the truth is, she is too excited to be sad. The future is finally happening.

She kisses him gently on each eyebrow and turns over on her side. Sleep comes quickly and pulls her far away.

Chapter 3

It is early and the city sky is still dark. The half-lit University buildings seem largely empty, the institute as well. Karen can hear her own footsteps. When she gets close, the security guy at the front desk looks up briefly, recognizes her and buzzes her in. She takes the elevator to the third floor and moves swiftly down the hall. The labs are empty and dimly lit. No one is around. It was like that in her previous labs, as well. No one else would come in as early as her, at least not consistently. She enjoys having this quiet morning time to herself. It is the best time of day to get serious work done. No disturbance, no competition for the shared equipment. Perfect.

She switches on the main overhead lights in the lab and heads straight for her desk. Sliding off her backpack, she glances, as always, at the picture of her and Bill above the desk. It is from a hiking trip a couple of years ago, one of their rare trips away. They are both beaming with happiness, surrounded by perfect autumnal colors. The picture makes her smile every morning. It stands out in the otherwise impersonal desk area filled with lab-books, notes and printouts of articles that she keeps meaning to read. A bit messy, but it aptly reflects work in motion, she thinks. She does not like desks that are too tidy, too controlled. The opposite, a piled-up desk with papers and data-sheets stratified into months and years of work, would also not work for her. She can't imagine it works for anyone; it is an affectation, really. A hint of the distracted genius, the scientist who comes up with the one experiment that says it all. No, science is hard work, she thinks, often interesting, sometimes inspired, but most of all, concentrated, hard work.

She has a few routine things to get started before she allows herself to go to the microscope. This takes some self-control, as this morning is more exciting than most. She is eager to find out what happened in the experiment that she set up last night. An extra 15 minutes of suspense before she looks at it won't hurt. First she should start the cultures for minipreps from her cloning. It was a pretty standard cloning and when she goes to pick up the plates, they look as they should. The positive plate has many more colonies that the control plate. Having her cloning work smoothly is the tiniest of victories. But still, she enjoys it, every time. Pick colonies now, she figures, and she can complete the preps and the next step this afternoon. She will be done one day faster.

She ducks into the tissue culture room halfway down the hall. First she switches off the blue overhead UV light. It makes the place look sterile, but whether it does the job, she's not sure. "Her" hood is the one furthest from the door, empty like the others. After removing her watch and ring, she washes hands thoroughly, switches off the UV light in her hood and starts the airflow. The little room immediately fills with a roaring noise, drowning out the fainter hisses and beeps of the incubators. The hood and the bottle that she removes from the fridge both get wiped off with 70 % alcohol. It will be washed and wiped again later. Worried about contaminating her unique cell lines, she is careful and does her most sensitive tissue culture work in the morning, before everyone else comes in to clutter the workspace and unsettle the air.

She finally allows herself to go to the microscope. There is a flutter in her stomach. She tries to tell herself not to expect too much, but doesn't manage terribly well. Hurrying down the empty corridor, she almost trips herself. She smiles, shakes her head and walks the rest of the way.

The microscope room is even smaller than the tissue culture room. It is quiet and dark. The only light is from the computer connected to the right-hand microscope. It shows the final frame from one of her overnight time-lapse recordings. It looks OK. One frame is not very informative, however, so she sits down and starts scrolling back through the images, the detailed recordings of last night's activities. Even after years of looking at cells, she still finds them fascinating. She captures all sorts of funny behavior in her videos. The cells stretch, round up and they even move around. Sometimes cells make new contacts with one another, a transient touch or a long-lasting grasp. Given the right stimulus, they will stick together and make larger, more complex structures. All of these events can be highlighted by multi-color labeling of the cells. In today's experiment, the fluorescent labels allow her to see the cells' internal architecture as well as their outlines. She watches internal cables being made, retract, being remade and the enveloping membrane shifting to accommodate the changing forces. Each structure has its own logic and function. Usually the path from observing to understanding what the cells are doing is long and slow. Occasionally, there is a significant shift, a moment when it all suddenly makes sense. But observing always comes first. And observing the cells is something she loves to do.

For each of last night's four videos, Karen finds the starting point. She instructed the computer controlling the microscope to come back to the same four places every 5 minutes and take a stack of pictures, all through night. The stack makes a box. Viewed in sequence, each set of stacks becomes a time-lapse 3D video. A video showing what each of her selected groups of cells were up to last night. She looks through them one by one. In the first frames of each video, she sees a faint, long line, a thin connection from one cell to another. This cell-to-cell connection, and what it may do, is what she is interested in. In one video, the connection breaks or regresses. But in the other three, the connections are retained. In these three cases, the cell that the connection comes from is very lively and appears to make some additional connections that she will have to look at more carefully later. The second cell, the one being contacted, remains healthy, in all three videos. The fluorescent cell death indicator she has put into the cells turns on in scattered cells in each of the recorded areas, but never in the pair of cells she is watching, the connected pair. Excellent. This is exactly what she had hoped for. "Good work, little ones" she says with a nod and a smile.

She has seen this twice before. But two examples could have been a fluke, so she set up several parallel recordings to get more examples. With this new data set, she feels convinced. This is really excellent. Now she just needs to capture enough events for a thorough analysis, so she can present it as a definitive result at the institute seminar next week. Although "just" an internal institute seminar, she wants to do well. And this result is exactly what she was hoping to include. One of today's videos looks clear enough to use in the talk, as visual illustration of what is happening. Or, more accurately, illustration of what she thinks is happening. She shouldn't get ahead of herself. But it does look very promising. She can set up another set of recordings tonight. And tomorrow, and maybe Sunday as well. Weekends are good. She can record both overnight and during the day, if no one else needs the microscope. This should allow her to squeeze in four or five more recordings before the seminar. She will need to analyze all the videos and do the statistics, of course. But the excitement of discovery is already there. In her gut, she knows she is on to something. It's there. She is all jittery. Before returning to the lab, she remembers to save the video files onto the lab server. Once, she forgot and the next person to use the microscope accidentally deleted her data. A mistake that will not be repeated. And, if one were needed, another good reason to get in early every day. No one gets to the microscope before her.

Happily buoyed by what she has just observed, Karen goes back to the tissue culture room to get on with the work that awaits her. Cells need to be checked, fed, or passaged into new culture dishes. Just routine work, but it is a pure pleasure now. Her thoughts return to the videos. The cells did what she had hoped they would. When they touch, they do not die. She thinks this is the role of the thin connections she has observed, keeping cells from dying. Other cells, well, some of them, they die. In fact, a fair number of noncontacted cells die under the stressful conditions that she has set up for this experiment. She knows she needs many examples to show that the proportion of cells that die is truly different whether they make contact or not. Later today she can count the number of dying cells in the videos she has so far and guesstimate how many more recordings will be needed for the statistics to be in order. If it holds up, this will be an important result in support of her idea. But even so, it will just be the first step, a correlation. She starts thinking of the next level, and what type of experiments would carry this forward. With a positive correlation, what she will need is evidence of causality. Seeing that the two events are linked, touching and surviving, does not prove that one thing causes the other. If she could show that touching causes cells to survive, then she would really have it nailed. As the fan of the tissue culture hood drones on, she goes through ideas for the next round of experiments, trying to come up with just the right one.

Two hours later, Karen has finished the cell culture work for the day. Her cells are all back in the warm and humid incubator. Those to be used for video-microscopy tonight are at the front. She leaves the hood running and goes back to the main lab to look for Etsuko.

Etsuko is at her desk, intently looking at the screen in front of her, playing and replaying a short video. To Karen, Etsuko seems tiny. The chin-length hair held back with colorful hair clips featuring flowers or animals adds to her girlish look, as do the frilly tops and skirts she sometimes wears. But Etsuko has a child of her own, possibly school age by now. So really she is more of an adult than Karen is. And, despite the superficial girlishness, Karen knows that Etsuko is a very serious scientist.

"It looks great" Karen says of the ever-looping video on the screen.

"Maybe. The new sensor I made. I am not sure it is quite correct."

Etsuko is usually very cautious. Karen knows Etsuko's tendency to endlessly retest and modify her tools sometimes irritates Tom. He does not always manage to control his impatience at group meetings, when Etsuko adds one too many preparatory caveat. "Just tell us what you see" he will insist. "And let's worry about the possible problems later."

"The hood is free now. I left it on for you." Karen says.

"Thank you. And thank you for letting me know" Etsuko nods slightly. "I should get started, before someone else takes it."

Yuqi is also at her desk, the one next to Etsuko's. Her long black hair is tied back in its usual ponytail. Etsuko and Yuqi have quite different features but they are both small, and wear similar clothes. Karen is slightly embarrassed by her continued reliance on their hair length as a visual cue for identification. She realizes that her poor recognition of Asian faces may be excused by her sheltered upbringing in a town populated by large, more-or-less blond Swedes. But it bothers her, nonetheless. The names are another challenge. Japanese names are not too difficult, but Karen knows that she fails miserably with Yuqi's name. "Good morning Ee-chi" her current best effort. Yuqi smiles. She does not seem bothered. Unlike some of her compatriots, she has not adopted a simple western name to circumvent the constant mangling of her name.

Since coming to the US during her PhD, Karen has come to realize how much she likes the mix of people from all over. Top US labs, like Tom's lab, are like the UN. One, maybe two, from everywhere. But, unlike the UN, this is not by decree. To Karen, this wonderful heterogeneity of science labs reflects appreciation of mind over body, curiosity over customs. That, and the fact that scientists can work anywhere. Labs are essentially the same all over the world. And scientists with ambition are usually willing to travel for the best opportunity. Despite everything, she has a lot in common with Etsuko and Yuqi.

Toward the end of the afternoon, there is a celebration in the lab. Chloe's paper got accepted in Nature, apparently. For any of the postdocs in the lab, getting a first-author paper in a journal like Nature is a big deal. It gives a huge push, career-wise, self-esteem-wise. So a celebration is in order. Karen knows this. She had two good first author papers from her PhD work and she still remembers how thrilled she was when they were accepted. None of hers were in Nature, though. And she has yet to have a paper from Tom's lab. Of course, she has not been here as long as Chloe has. And she reminds herself that she is well on her way to something now, with this morning's results.

Getting papers published in the top journals also matters to the PIs, the heads of lab like Tom. But Tom has many such papers to his name by now, so he can afford to be generous with his praise today. He tells the lab that not only did Chloe do all the terrific work in the paper, but all the ideas were hers as well. He adds jokingly that he is just along for the ride, to pay the bills. Karen finds this self-deprecation a bit excessive. It would seem like false praise at home, maybe a bit mocking. She is fairly certain that Tom has had more significant input than he lets on. Chloe's paper is a natural extension of earlier work from Tom's lab, on the Myc oncogene. Doesn't he feel false, then, talking like that? Sometimes she is not sure she can read Americans correctly. Karen knows from lab meetings that Chloe has done an amazing amount of work on this project. So she has probably earned the paper. Of course she has. But maybe Tom is overdoing it, just a bit.

After the toast, people stand around for a while, asking Chloe questions and congratulating her. Vikram is being silly, as usual. He is a funny one, Vikram. He likes to give the air of not taking things seriously. But, behind the fun, there is plenty of drive, Karen knows. He is one of the lab's night-owls, but several times she has met him, still working, in the morning when she comes in. And when he talks about his science, he is serious, focused. He seems both knowledgeable and smart. Whenever the release button is pushed, however, he falls effortlessly back on joking. Karen does not know exactly what his background is, but it must have to do with England, given the accent. In addition to Vikram, Michel and Juan are also hanging around Chloe, paying her lots of attention. Flirting, it seems. How does she do it? Karen wonders. Physical attractiveness is an important part, no doubt. But it is not just that. There is also Chloe's easy charm, charisma even. That will certainly be a significant asset on the US job market, Karen thinks. And a Nature paper to top it off—a winning combination.

Karen puts down her glass, having only sipped at it. She normally likes champagne and she did notice that Chloe had bought the real thing. No sparkling German Sekt when it is a Nature paper. But there is work to do. She heads back to the lab, reconnecting to the excitement of her results earlier in the day along the way. Her work is more original, she thinks. It is far from what anyone else is doing in the lab. And it is starting to pay off. But, she needs to think hard about the next round of experiments, about how to prove causality. That will be the finishing touch for a major paper. Back at her bench, she runs through what she needs to do for the rest of the day. She suddenly remembers that she never got around to congratulating Chloe. But everyone else did, so it is probably OK. It is OK. Right now, she needs to set up a new round of recordings. Her calculations after lunch showed her that at least five more positive experiments are required to get convincing statistic significance. More would be better. So she will push through whatever extra experiments she can manage before next week. With this new observation, her next institute seminar will not just be "nice" and "promising", Tom's rather cautious word of praise after the last one. This time, she will knock their socks off.

Chapter 4

"I have to go in to work for a little while." Karen says, apologetically.

"Today?" asks Bill. "It's Saturday. And Erik is coming, you know, and Ashok, for dinner."

"I'm sorry. I know we decided to try to keep weekends work-free."

"Given your twelve-thirteen hour days during the week, it's pretty reasonable, don't you think?"

"Of course. It's just that I can use the microscope in the Küstner lab over the weekend. You know, the one with the laser cutter, for my special experiment. I have to work around their schedule."

"I know, you told me. But it's the third weekend in a row." He has that sad, disappointed look. She almost can't bear it. His deep brown eyes, the warmest, gentlest eyes, looking at her like that. He deserves so much better, her most wonderful husband.

"It will just take a couple of hours, I promise. We can get everything ready for tonight beforehand. I'll clean while you shop. And if I take the car, that saves an hour in transport." Bargaining and pacifying in one. She knows that Bill hates cleaning.

"Well, OK, but only if you are sure you will be done in time to pick up Erik from the airport at two."

She is not sure. She will have to leave the car with Bill. Busses are less frequent on the weekend, so it could take her an extra hour to get back.

"Sweetie, I'm sorry. I just need to do this, I do. Can you pick him up?"

Erik is her friend and she knows it is not quite fair to ask Bill to take over. But he agrees. He deserves a real thank-you, she knows. She holds out her hand to him, cocking her head playfully. They still have some time for Saturday morning friendliness.

On weekends, the lab feels very different. People come in when it suits them and have a more relaxed air about them. Karen also has special rules for weekend work. She allows herself to do only what she feels like doing. These days this includes routine work in the tissue culture room. Getting it done now will give her a head start on next week's work. She switches off the UV light, turns on the airflow and takes out media to warm up. She does have work to do on the microscope in the Küstner lab as she told Bill, but since she is here, she might as well take care of some other things as well. There is so much to do. Ever since the promising results a couple of weeks ago and the well-received institute seminar, she has wanted to be in the lab all the time, setting up more experiments.

Before settling in to do cell culture work, she goes down to the second floor to check on the microscope. Karen was very happy when Bettina Küstner, possibly helped along by friendly nudge from Tom, agreed to let her use their microscope. Greg, the engineer who built the laser cutter and much else on this microscope, spent many hours, early on, showing her how to use it. He probably wanted to make sure his precious machine would be in good hands. Having gained his trust, Karen is now allowed to use the microscope on her own. The incubator box needs to be on for at least an hour to warm up before she can get started. It is good that she was realistic with Bill about her timing. There is no way she will be done in time to make the 2 P.M. pick-up. This experiment is difficult and important and she needs all the microscope time she can get, visitor or not.

Karen lets herself into the half-lit Küstner lab. Strange, she thinks, how this lab is always empty on weekends. Yet it is a successful lab. Bettina Küstner is very highly respected. Her lab has developed some pretty cool microscope techniques, she remembers from institute seminars. Maybe the difference in work habits reflects that the Küstner lab is full of engineers and physicists. Most of their work is on a computer. No need for long hours of doing and redoing tedious bench work. Maybe they just need a few good hours each day. Well, she thinks, computer-work can also be done from anywhere. So maybe some of them are working right now, from home. This thought satisfies her, for some reason. As she enters the microscope room, her attention turns to more practical matters of dealing with the delicate machine. She needs to be vigilant. The microscope is nontrivial to use with all its custom-made devices. She is also still optimizing the experimental setup, so she needs to pay attention to all the possible parameters. Once she has the experiment working well, she will be able to definitively test her ideas. Exciting—and a bit scary as well. That is the risk with a truly decisive experiment. It may turn out great and support her ideas. Or it may prove her wrong. 'Hero or goat' she thinks, echoing a lab joke, whose origin she does not really understand.

When Karen finally gets home, it is almost evening and she has rehearsed her apologies. But lively voices and laughter are coming from the kitchen. It seems Bill and Erik are getting along just fine.

"Ah—there she is. The lab rat returns." Teasing is a good sign, so she may be safe from reproach. She gives Bill a kiss, puts down her bag and turns to Erik. He has an opened beer can in his hand. She hesitates, uncertain about how to greet him, but decides to ignore her instinctual Scandinavian reserve and gives him a slightly clumsy hug. Erik's familiar face reminds her of many old attachments and of a world she suddenly misses with an unexpected acuteness. The hug is for all that. The bewildering feeling is gone almost as quickly as it arose. But Erik appears pleased to be welcomed in this way, clumsy or not. So it seems she chose well.

"Erik, it is so good to see you. How are things back home? How is the lab? How was the meeting?"

Too many questions, but Erik seems unfazed by her eagerness. Bill will think that she is trying to make up for being late. Erik might or might not understand. He starts updating her on the lab, Nils' lab, back in Sweden. He describes his own sub-group and their projects. He talks a bit about the field and what was presented at the meeting. She struggles to stay attentive but nods along and injects an occasional simple question. Apart from the unexpected flutter a moment ago, Nils' lab seems so far away, so unimportant now. When Karen was starting out as a PhD student, it was her whole world. Then she visited a collaborator's lab here in Boston, for about a year. This was a revelation to her. Here, no one thought there was anything wrong with showing ambition. She was immediately accepted, and even appreciated, for her ability and effort, for all her hard work. No quiet hostility, no "Law of Jante". What a relief that was. Now she can enjoy this acceptance full time, in her adopted country.

Erik is slowing down. She should say something.

"I have to admit I haven't been keeping up to date with the field. A lot happening, it seems. And I guess Ulla must have finished her PhD, and Torsten as well. Do you know where they have gone off to?" Unexpectedly, Erik does not pick up on the invitation to an easy chat about other former PhD students.

"The field is getting really tough, and nastier too. It is almost impossible to get anything published in a decent journal. At least for me, it is. 'Not enough new insight' or whatever they usually say. Journal editors are so superficial. What do they know?" He sounds bitter. "It's not like what they publish from other labs is any better. It was easier when Nils was the corresponding author. The grand old man."

Karen wants to avoid going too deep into this. The truth is that she has kept herself updated on her former field and sees that Erik's and Nils' work, all the work from her old department, is getting a bit dated. But she most definitely does not want to say this to Erik. She does not want to hurt his feelings. He took the safe option of staying on in the old lab and rising through the ranks. He did both his PhD and postdoc with Nils and he is now associate professor there. Nils obviously thinks very highly of him. But this also means that Erik has stayed close to Nils' original vesicle work. More cellular genetics have been mixed in, but no major innovations. Erik has not strayed from the path and he sees only the trampled ground. She cannot avoid thinking this.

Trying for a change of topic, Karen turns to Bill. "Anything I can help with, my dearest cook?" He shakes his head. It is starting to smell very good. She is hungry and she knows that some of her favorite middle-eastern vegetarian dishes are on their way.

"Sorry." says Erik. "I shouldn't complain. I guess being at the meeting got me worried. There's just so much going on that it is hard to keep up, even within your specialty. And lots of people are doing things similar to what I had planned to do. So, well. . ."

"I haven't been to meetings for a while, I have to admit." Karen looks at Erik again. "I don't have a good story to present yet. And no publications from my postdoc yet either."

"But at least you are in Thomas Palmer's lab. That must be a great environment. And there are so many possibilities in the US."

"Well, yes. It is certainly a good opportunity. But you still have to make your own way. There is no army of technicians to help out. And Tom doesn't design the projects, or steer them. We postdocs have to build our own fires. You know?" It probably sounds as if she is complaining now, or competing with Erik for worries. This was not her intention. "Of course I wouldn't have it any other way." She adds quickly. "Anyway, it is good to a have a husband with a steady job." She sees Bill give her a puzzled look. Perhaps that was too flippant. Well, she knows it was.

"Whatever you do, I'm sure it will be good. And you'll publish well when you are ready." Erik says diplomatically.

"I hope so. Right now, I'm just trying to do something interesting." She finds herself surprisingly reluctant to talk about her project. She opens the fridge and pours herself a glass of the opened wine.

"Maybe I should set the table?" She suggests to Bill's back. "Ashok should be here soon." She turns to Erik and adds "You will like Ashok. He is a really interesting guy, a colleague of Bill's."

As if on cue, the buzzer goes. "That must be him now. I didn't realize it was so late." She hurries off to answer the door.

Karen introduces them. Ashok, a short wiry Indian with lively dark eyes, is quick to extend a surprisingly large and hairy hand to the tall, pale and slightly hesitant Swede. A few questions and answers and they have placed each other, their jobs and relationships to Karen and Bill. Karen fetches Ashok a glass of wine, knowing that he prefers this to beer. When she returns, Ashok has already gotten around to asking Erik about his family background. He seems to find all life trajectories interesting, mapping and analyzing them as if for some grand project. Perhaps there is such a project. She is not sure. But at least it makes for easy conversation with visitors. She is more than happy for Ashok to have hijacked the conversation.

Karen goes back to the kitchen to see if Bill needs help. It turns out he doesn't, but she lingers in the kitchen anyway and gives him an extra hug. She asks quietly about his impression of Erik. Instead of answering, he offers her tastes of the various dishes. Bill is a gentleman and prefers not to indulge what he considers her occasional inclination to be judgmental. She gets it. She starts bringing the colorful plates and fragrant dishes to the dinner table instead.

Bill's variations on middle-eastern delicacies are duly admired and, over the next couple of hours, they slowly work their way through them. Karen occasionally retreats to the kitchen to warm more flatbreads. But she is back quickly, not wanting to miss any of the conversation. Ashok has turned them toward one of his favorite subjects, the nature of scientific inquiry.

"You know, you have two real scientists here," Bill warns him with a friendly smile, "not counting me. So be careful what you say." To Erik he explains. "Ashok knows a lot more about science than we do. He actually thinks about it. He teaches a very interesting course on philosophy of science. Probably one of the most challenging courses we offer to our seniors."

"You flatter me, Bill. It is certainly not the most popular course. But I do love science, the study of science."

"But you did not go into science yourself?" Erik asks.

"Not into experimental science. I leave that to the real doers, like you and Karen. No, what I am fascinated by is what you might call the process of acquiring knowledge, in particular the scientific method. It works in such a surprising way, so unexpected."

"Why unexpected?"

"Well, it is naïve, really, isn't it? Naïve in the sense of starting out by saying I don't know anything about the world except for a few logical formalisms. If we leave out the theoretical sciences like math, then grasping the world with the scientific method is sort of naïve and blundering." Ashok is primarily addressing Erik. But Erik looks unconvinced. This does not deter Ashok.

"To understand the natural world, one must start with observations. And then build on this with inferences and hypotheses. You see things happen in a particular order. Like you kick a football at an angle and it flies forward to the right. Multiple events like this allow you to expect certain outcomes and make hypotheses about causality."

"But why is that naïve?"

"Patience. I'm getting there. Everyone, be they laymen, scientists or religious fanatics, operates with causality, explaining why things happen. The important question is what makes you accept an explanation as true, and following that, how you go beyond simple cause-and-effect to describe the natural world. Do you build an elaborate construction that carries explanations for everything? Religions all do this, of course."

"Religions are fairy tales." Erik states, firmly.

"Maybe. The point is, there may be an endless number of explanations, which Gods, how, when, and so on. The alternative is to use the scientific approach: start with not knowing, then observe, measure and question. Finally, test any inferences or expectations by experiments. And respect the answers. But you have to accept that you learn only as much as the experiments reveal directly, never more. And this is the hard part, because then you have to admit that you know very little."

"I think most people would say that because of science, we know a lot." Erik says.

"Whether we know a lot or a little is hard to say, I guess. But I would separate out the technological part from any deeper knowledge. We often recognize cause and effect well enough to use it effectively even if we don't really understand why or how it works. But this is not how we are taught to think about the world. Have you read Wolpert? His book, "The unnatural nature of science" is really fascinating on this topic. We teach students about science in the classroom as though it is like any other information that you need to accumulate. So they do not know the difference."

"The difference between what?"

"The difference between knowing something because you are told it is true and knowing something because you are convinced by the proof. Take DNA. By now, most people know that DNA can be used to identify and characterize people. The better-educated ones may even have some idea of how it works. But if you ask them whether they know about DNA in a different way than they know about the constitution, the presidents or the value of democracy, they will not be able to explain."

"I disagree." says Karen, "I think that people do know the difference. They would say that DNA belongs to science and its role can be proven. As for political views or even issues of faith, there are some facts, plenty of opinions, but no proof. People know that there is a difference between being told an explanation is correct and it being proven. Even if not everyone can evaluate the scientific evidence directly."

"OK, maybe. But my point is this: do they understand what scientific proof actually consists of?"

"Evidence, experiments." Says Erik.

"Yes, but experiments can usually only disprove a scientific hypothesis, and should be designed with that purpose." Ashok is clearly enjoying himself, leading them to this. "Of course you know that, as scientists. And you know that disproving one thing is not the same as proving the opposite. There can be valid explanations that you have not thought of. Accepted scientific explanations are often proven wrong later on. Given the nature of the approach, this makes sense. But it also means that science does not give you the final explanation. It gives you one possible interpretation that, currently, is not inconsistent with the available data. You know of Occam's razor, of course."

"The simplest possible explanation is to be preferred." Karen answers, dutifully.

"The one that needs the fewest assumptions, more precisely. You have to address that one first. Can you disprove it or is it consistent with the observations? But even if you cannot disprove an explanation and therefore have to embrace it, it may turn out to be incorrect."

"I suppose," Erik concedes. "But that is all so theoretical. And it makes science seem so arbitrary. As if we have very little knowledge and what we have won't stand the test of time. But that is not true."

"It doesn't seem that way, I agree. But that is in part because we have so much data about the world. So our current interpretation of the world is highly predictive. Of course, biology has added complications. I've enjoyed learning about this from Bill."

Bill nods in recognition of these conversations, adding his brief summary: "Too many variable parts, too many unknown factors."

"Yes, factors you cannot possibly fully predict." Ashok pauses and continues in a different, slightly teasing tone. "It must be difficult, having to rely on statistics to tell you that you are not very likely to be wrong."

"We survive." Karen responds, with mock-sarcasm. She and Ashok have discussed statistics before. They have agreed to disagree.

"But it worries me, it does." Ashok remains serious. "It worries me that everyone uses statistics to test things and then think they have been given an answer. It doesn't tell you that your interpretation is correct."

"Be fair here." Karen protests. "We know that. Statistics is necessary for all kinds of reasons. For one thing, so that you know not to be fooled by random favorable events into thinking you have an effect."

"Well, maybe you know. But I am not convinced about scientists in general. And more broadly, it seems that many scientists know very little about how the scientific method works and its limitations. It's a bit shocking that you lot are allowed to practice your profession without a proper theoretical foundation." He smiles mischievously.

Erik still looks puzzled. He clearly wants to say something, but hesitates. After a moment, he turns to Ashok.

"Well, all this does not sound very satisfactory, does it? Nothing is ever proven and statistics are misused. But then why are you so fond of the scientific method?"

"Sorry, forgive the digression. But yes, I am fond of it. For two reasons, really. One is the respect for nature and the universe. You do not presume to know. I find this way of approaching the world both liberating and humble, especially when compared to the many perverse human endeavors where some group presumes to know what is right and best for everyone."

"I think a lot of people would say that about scientists."

"Individual scientists, perhaps. But the method, no. As I was saying earlier, when you examine something scientifically, you always start by acknowledging that you do not know the answers. You observe, and if things appear to be causally connected, you make a hypothesis about how they are connected. But that is only step one. Then you have to test your ideas by setting up experiments that could disprove them. And you have to respect the results of the experiments, even if they do not fit with your ideas. You are obliged to accept it when the data tell you that you are wrong. Such a result is also a step forward. You have to be brave and embrace it."

"I am not sure that most of us are that brave." admits Erik. "We are not exactly challenging the current world view with our little experiments. It is more like we are coloring in the details."

"Well, until you find something unexpected, right?" Ashok grins, "The expected results are boring, no?"

"Ah, this is when it becomes clear that you are not an experimentalist." Karen adds with evident satisfaction. "Getting a long-hoped for result can be really fantastic, making all the hard work worthwhile. See, you have an idea about how something works. Maybe you have seen a connection that no one else has seen. Then you set out to test your idea—the suspense time—are you right or not? If it turns out that you are right, that you have found the explanation for how a part of the natural world works, it is amazingly satisfying." Ashok tries to break in, but Karen anticipates his objection. "OK, so you will say that what you have is one explanation that is consistent with the facts, not necessarily the right one. But now your idea is no longer just an idea, it is supported by experimental evidence. Getting the idea and providing the experimental support is fantastic." She gives the "and" a strong emphasis. "It is so much better than reading about it. This is what makes doing science so special." She hurries on. "The other, hugely satisfying part is designing great experiments. Some of the best "Aha!" moments are when you realize you have come up with "The" experiment. An experiment that is doable, elegant and critically tests your idea. This, to me, is amazingly gratifying." She pauses, and adds, as an afterthought. "To be honest, if it wasn't for the possibility of those moments, working in the lab would be awfully tedious."

"But you are always going back to the lab!" Bill protests.

"Well, that is because I'm running my own experiments, testing my ideas. It's not just the doing. In fact, I really hated working as a technician for the few months I did that." She pauses. "Sorry, I guess I got a little carried away there." She turns to Ashok. "So, what is your second reason for being fond of science, or rather, the scientific method?"

"Well, it's related, really. The possibility of the unexpected. Since we do not know much, there is always a possibility of uncovering something completely new. If you already have your answers, your faith, there is nowhere to go."

"I think we can agree on that, in principle." Karen says, "In truth, it is rare that anyone strays far from the paradigm of the day. Fortunately, small discoveries have their place, as well."

"I think that most big discoveries are triggered by random observations, rather than an unexpected outcome of testing hypotheses," says Erik. "I mean, think of how antibiotics were found. It is a fantastic story. And how a drug like cisplatin was found. It is now widely used to treat cancer."

"But such observations would fall into this category, as well, would they not? The unexpected." Ashok asks.

"I suppose. But in real life, in the lab, there are so many little things that do not fit neatly with what you expect. Most of these are due to human error, to variation, whatever. It is hard to know which ill-fitting and unexpected result to take seriously. Of course, we would all like to find something like that, discover the next antibiotic or knockdown by RNAi." Erik adds, with more regret than hope, it seems.

"Enter the mighty ruler of experimental science" Karen says, and pauses for attention. "Repetition. Repetition. Once you notice, can you repeat it? It gets back to statistics, of course. But it is also a fundamental tool, a light in our darkness." She smiles teasingly at Ashok and then continues more seriously, "But you are right about respecting the data, the experimental results. That you cannot alter or misrepresent data is an absolute dictum. It can be hard to accept an unwelcome result. It requires self-discipline. But, if you don't, then you are not doing science. You could say this respect is sacred."

"So no sacred truths, but a sacred method?" Ashok suggests.

"Something like that." Karen seems distracted all of the sudden. Ashok prods her gently.

"May I venture to guess that you have experiments going on that critically test a hypothesis of yours?"

"Yes, I confess. I can't help thinking about it."

"I'd love to hear about it. But keep it simple for an old generalist. No fancy modern terms. I won't understand them."

"OK. Well, the concept is actually quite simple." Karen starts "I am trying to understand how cells decide whether to stay alive or to die, to commit suicide, when they are together with a lot of other cells, like in a normal tissue. Secreted factors can control this. These are molecules that simply float around, made by some cells to influence many other cells. Don't worry, I won't go into details," she sends a quick smile to Ashok. Bill has started clearing the table. He knows this part.

"I am a cell biologist by training, in part his fault," Karen smiles and nods toward Erik, "so I decided to look for differences in cell behavior between cells that die and those that do not when I make small pieces of artificial tissue. Because of the way I labeled the cells, I noticed some very thin, very long connections between some of them. And then, by making a bunch of video recordings, I found that the cells that had these connections survive much better than other cells. Connections that allow cells to touch other cells far beyond their immediate neighbors, that is."

"What are these "connections"?" Erik asks, the quotation marks apparent.

"Well, a new kind, as far as I know, not described before. They are very thin. By now, I know some of the details of how these connections are made, what they are made of, and their dynamics."

"Wow. Cool." Erik says, spontaneously.

"It is. Very cool, actually. Anyway, I have statistically sound evidence for the correlation. When cells have the connections, they don't die. But many of the nearby cells without connections, they do die. I know enough about the molecules involved to make the story acceptable in a modern cell biology way. But I still need to show that the thin connections, as such, are needed for the cells to survive."

"So your key experiment will address the issue of causality directly?" Ashok says, understanding.

"Yes, exactly. What I am trying to do is to cut the connections with a very fine laser and then see whether the cells behave differently afterwards. Do they start dying like the other cells? Of course I have to do this very gently, so the cells do not bleed from the cut, so to speak."

"And if the cells behave the same after you have cut the connections, that would disprove your hypothesis?"

"Exactly. It is a classical setup. I can disprove my hypothesis very cleanly. Or I can get support for my hypothesis. I can't get positive proof for it, of course. There could be other reasons why the cells die if connections are severed. But if I can make this experiment work and if it shows what I hope it will, well, it would be really great. So I guess you understand why I say the expected result can be most welcome. It would give me an extremely compelling story."

"So you must be really excited about doing the cutting experiments?"

"Yes," Karen laughs. "Excited and terrified, alternating. It is technically difficult and I haven't quite got the set-up tuned correctly yet. But I'm trying hard." She pauses, "The point is, if I disprove my hypothesis, if the cells do not care whether the connectors are cut, I am done for. Then the project is dead. I may in a formal, theoretical sense have made a step forward. But in real-life terms it would mean that I have nothing to publish. I would be back where I started, with almost 3 years of work wasted. No publication, no job prospects. So yes, it's terrifying. The reality of risky projects." Despite the somber final comments, she finishes with a determined smile.

"That sounds interesting, Karen." Erik says, in a slightly cooler tone. "I am curious about these structures and what molecules might be involved." After a quick glance in Ashok's direction, he adds. "Maybe you can tell me about it tomorrow morning?"

"Sure" Karen obliges, but with slight unease. She reminds herself that Erik is not on his way to the cell biology meeting and 10,000 eager scientists, but on his way home. And he is a friend. Still, none of this is published.

"Turkish coffee, anyone?" Bill comes back with the elegant copper pot, matching tiny coffee cups and a plate full of fantastic-looking flaky baklava. It is a good occasion to switch topics.

Chapter 5

"Chloe, Tom wants to see you in his office."

Deidre, Tom's secretary, is standing at the end of Chloe's and Juan's bay, but is keeping a safe distance from the workbenches.

"Do you know what it's about?"

"Nah, he didn't say. But it sounded important."

Chloe gets up quickly and follows Deidre back out of the lab. She will never understand why Deidre is so afraid of what goes on in the lab, but feels she can well afford the kindness of not keeping her there unnecessarily.

When Chloe gets to the office, she sees Tom at the round table with a neat-looking man, probably in his late thirties, sporting dark brown hair and a well-trimmed beard. He is wearing an off-white linen shirt, ironed but no tie, and making notes on an old-fashioned notepad. So probably not a scientist and for sure not an admin person, Chloe thinks. They stand up as she enters.

"There she is." Tom beams pleasantly. "Chloe, this is Frank Lockwood, science reporter from the New York Times. He is here to do a story on your

Nature paper." Tom turns back to his visitor. "By the way, although the paper was accepted a month ago, it won't be out until early January. I assume you know about the embargo date and all that?"

"Yes, naturally." Frank answers. His voice is calm and pleasant. "My story will run the same week, right after the embargo date." He holds out his hand and steps toward Chloe. He smiles with what appears to be real warmth and delivers a brisk handshake. "Pleased to meet you."

"Likewise."

"I've just told Frank here that this is your story" Tom says "so rather than spending all his time with me, he should talk to you. I've given a bit of background about the lab. But you should tell him about the paper."

"Sure. I'd be happy to." The New York Times has a serious science section, she remembers. So Frank must be one of the good guys, a science reporter who understands and appreciates science. Tom is obviously ready to get onto other things and remains standing. Chloe takes the hint and leads Frank off to the small meeting room down the hall. Leaving him there, she hurries back to her desk to pick up her laptop.

When she returns, Frank is studying his notes. He looks up quickly with an expression of serious and keen intent. Perfect, she thinks, he is not disappointed that he is no longer talking to the famous professor. Since it is just the two of them, she sits down next to him and positions the laptop so they can both see the screen. As she opens her Powerpoint presentation, Frank gestures for her to wait.

"Chloe, I really appreciate you taking the time to talk to me. But maybe I should tell you what I am after, to make our discussion as fruitful as possible." Chloe realizes that he probably knows scientists' habits and is trying to forestall 45 minutes of a full-throttle research seminar. "I've read your manuscript. Dr. Palmer, Tom, sent me a copy. In confidence, of course. What I'd like now is your angle on the discoveries, the story in your words. Tell me how you found Jmjd10, how you figured out that it acts via Myc and what you think it means for cancer cells. And more on the drug prospects, perhaps." He has come well prepared, she realizes. Excellent.

"Science anecdotes, if any, are welcome. And if it's OK, I will interrupt with questions along the way, maybe try out some rewordings. I need to make it understandable to the lay reader." He stops briefly. "By the way, beautiful work. My PhD was in cancer biology, so this is right up my alley. Or was. It's been 8 years since I was deep into it myself. I'm really impressed." A pause. "So, with all this work in one paper, just tell me whatever you think is important. Then I'll select what I think conveys both the science and the human side of your discovery. A half-page article can only hold so much. I'll get a draft version of my article to you and Dr. Palmer for feedback before anything goes to print."

Thus reassured, Chloe begins talking, without using her slides.

"So, I started out looking for genes that would regulate tumor growth."

"Let's start with that. A lot of genes affecting cell growth and cancer have been found already. Why were you looking for more?"

She hesitates, slightly taken aback. She did not expect the premise of her work to be questioned outright. But it is a reasonable question, after all, so she answers as straightforwardly as she can.

"To be honest, my hope was simply to find something new, something with potential, so I could develop my own little patch to work on. But I also had a particular idea about how to look for regulators that I wanted to explore."

Frank nods.

"I set up a screen to examine all protein-coding genes in the human genome, knocking down their expression and looking for specific effects on tumor cell survival and growth."

"OK. I'll need to get the enormous scale of this effort across to my readers, that you looked at around 20,000 genes."

"Well, I didn't do them all one by one. I used pools first, each targeting twenty genes. Still, it was a huge endeavor, over a thousand pools to look through. At the time it seemed endless. Anyway, as you say, lots of tumor regulators have been identified already. Using a knock-down screen is not new, either, but my idea was to test growth under conditions similar to what tumors experience in our bodies. So I'd grow the cells in a tissue-like soft three-dimensional, or 3D, gel instead of the flat, hard surface of a plastic dish. Cells behave very differently under these quite dissimilar physical conditions. So not using plastic dishes seemed like a good place to start."

"That sounds reasonable. Why haven't other researchers done this?"

"Because regular plastic dishes are so easy to work with. You just pull them out of the bag and the cells behave, every time. Doing experiments in 3D gels is much harder and more variable." She explains in more detail and adds. "The other thing I did differently was to look directly for cells dying by apoptosis, not just for less growth."

"Cell death, cellular suicide."

"Yes, because that's what you want. You want to trick tumor cells into dying. It was more work to set up the method, but it turned out to be easier to pick out the promising pools once I got started, because they were brightly labeled by the cell death marker."

"Right. I think I'll be able to describe the logic of your approach. And discreetly convey that it was a lot of work, doing all this in a non-standard way."

"A lot of work, absolutely. I basically spent 2 years of my life doing almost nothing but running the assay again and again." She shakes her head with a slight smile, remembering. "I did have preliminary hints, though, that I could identify new genes this way. Otherwise it would have been crazy."

"So how many hits did you get? And was it easy to choose which one to focus on? Did the one you chose ..." he looks at his notepad "did Jmjd10 stand out in some way?"

Chloe laughs, "I call it JimD10 most of the time, it's easier."

Frank seems to appreciate the nickname and writes it down. She explains how she identified JimD10 and what she learned about it.

"So," she concludes, "the upshot is that JimD10 is only needed when tumor cells grow in 3D gels. The same cells growing on the flat tissue culture dishes don't care. This was really intriguing. It also showed me that my idea of setting up the screen the way I did was a good one. And, what was super cool was that normal cells—I mean cells that don't make tumors—don't seem to need JimD10 either." She grins, remembering.

"Aha. So then you knew you had a perfect target, a gene that tumor cells need but that normal cells can do without. Great."

"Yes. Exactly. I also looked at what happened in the mouse. Luckily there was a company that had made JimD10 mutant mice. The mice were reasonably normal, but looked like they were suffering from mild radiation poisoning. You know, problems with the tissues where cells are being replaced all the time, like the gut, the hair, blood. We can talk more about that, but the most important part was whether these mice would be more or less likely to get tumors."

"Yes, let just focus on the tumors for now."

"OK. So, we can use chemical carcinogens to induce tumors in the skin of mice. It is a bit nasty, but it is a very reliable test. The JimD10 mutant mice got essentially no tumors. This was amazing." She locks her eyes on his face, making sure he is paying attention. "Just by having less JimD10, the mice were protected from getting tumors. That was when I knew I had it." A flicker of delight crosses Chloe's face.

"So, let me recap." Frank says. "You found a gene that is specifically needed for tumor cells to survive and to grow in 3D gels. Removing the gene makes mice resistant to carcinogens, making a strong case for relevance to human cancer."

"And it's an enzyme, don't forget." Chloe interjects.

"Yes, of course. I'll explain that drugs are often small molecules that prevent enzymes from working. So JimD10 could be an excellent drug target. And you actually identified small molecule inhibitors of JimD10, as I recall."

"Yes, I did, or, we did. But you're jumping ahead, I need to explain how I figured what JimD10 actually does." A measure of pride is creeping into Chloe's voice. This is important.

Frank knows he is unlikely to use most of what is coming next. She wants to go deeper into the basic biology, the part that was intellectually exciting for her. There will be too much detail for his article, but some of it might be useful. So he indulges her, remembering his own inner geek. And she tells a compelling story. She describes the analytical journey that she took to understand how JimD10 works, joining numerous observations and designing clever experimental tests of her inferences. Like advanced detective-work. He asks a few questions along the way. When they get to JimD10 acting on the well-known oncogene Myc, Frank starts taking notes again.

"So JimD10 works on Myc. This must've been gratifying for Tom, having a molecule he has worked on for so long take center stage again."

"Yes, Tom got very excited about the project at that point." She pauses.

"Right. So let's get back to the small molecule inhibitors." Frank prompts.

"Of course we realized the potential of manipulating an enzyme needed for Myc activity: An inhibitor of JimD10 might be a good cancer drug. A chemist at the University, Kumar Singh, had been developing small molecules to inhibit enzymes like JimD10. So we arranged to test his compounds. One of them turned out to be an excellent inhibitor of JimD10. And in my cell assays, the effect of this inhibitor was exactly as I had hoped." She smiles, but then sighs.

"Once we included the inhibitor in the paper, the reviewers wanted more. They wanted us to test it in mice with genetically induced tumors. This seemed a bit over the top. I mean, I had inhibitor results with tumor explants and I had shown that the JimD10 mutants were resistant to carcinogens. So it was completely clear that JimD10 was needed for tumor growth. I was also worried another group was going to get the story before us."

Frank scribbles away, appreciating this real life flavor of doing top-level science. The last, painful push it often takes to get an already very substantial piece of work into a top journal.

Chloe continues. "The reviewers wanted us to test a specific kind of tumor, to make the Myc connection stronger. Luckily we had the mice we needed. Actually, they were made in Tom's lab originally, so. . ."

Chloe trails off for a moment. "Anyway, it all worked out in the end." She shrugs and flashes an ironic smile "And, it turns out I need not have worried. The competitors were apparently not even close to publishing."

"So," Frank says, slowly, looking at his notes "would it be fair to say, based on your mouse experiments, that you are optimistic that this inhibitor can cause pre-existing tumors, like in patients, to regress?"

Chloe nods, starting to answer "That's what I hope, but you know it is a long way from a drug in the lab to a drug for patients. I'm. . ."

Frank's cell phone buzzes. He gives Chloe an apologetic look and picks up. "Yes. No. That's fine. We will be right there."

"That's the photographer. He is here now. I think I've got what I need, so maybe we should move on? I'd like to get a picture of you working in the lab to go along with the article."

"Right." Chloe says, fighting off disappointment. Frank notices.

"Why don't I take you out for a coffee after the photo-session?" he says "I would love to hear more—also about how you plan to pursue all this. Off the record, of course."

Chloe agrees easily.

They head toward Tom's office and the waiting photographer.

They get their coffees, one latte, one Americano, and find a quiet table toward the back of the café.

"So where does the work go from here?" Frank asks. She starts to answer, then hesitates. "Don't worry. It's safe." He says, with a short laugh. "It's not like I'll run off and do experiments in my spare time. I'm just interested. And it won't go into the article. I have more than enough material already."

"Well, I guess I'd better get used to it. I'll be telling lots of highly competitive scientists about my ideas and plans when I go on job interviews soon. So now is not the time for being secretive." She can't help but be a bit worried about the upcoming public exposure of her most precious ideas. Who wouldn't be? Someone might just get a little bit too inspired. But Frank is right, she need not worry about him.

After half an hour of identifying interesting questions from her work and describing ideas for how to address them, Chloe pauses, sips her now cold latte and laughs. "I could go on for ever. I am probably boring you." She tilts her head, touches his forearm lightly for emphasis.

"No, not at all. I've enjoyed a dip back into the whirlwind of pushing new ideas around. Your enthusiasm is infectious." Frank looks at his notebook briefly. "Tell me about yourself, your background."

"My background? Not for the article, is it?" she says, guarded again.

"Not if you don't want it to be. I'm just curious."

"But why do you ask?"

"Well, I detect a slight accent, but I can't place it. So, origin unknown. And dedicated people interest me. I wonder where it all comes from, the interest and the drive. Are you from a family of scientists?"

"No, no. My father is just an ordinary doctor. In Heidelberg."

"So you are German?"

"Well, yes. I am a German citizen. But my father is Hungarian." She pauses. "My mother is German. And I grew up there. I went to an international school for some time and I came to the US for my PhD."

"That explains the almost absent accent."

"I suppose. I've been used to different languages from early on. German, Hungarian, French, a bit of Italian. My father also spoke English to me. I asked him to stop when I realized how bad his accent was. I guess I was a bit of a brat, being ashamed of his accent in front of my school friends." She flashes an apologetic smile. "He learned English when he came to the US in'56 to do an undergraduate degree. But he didn't get into medical school in the US. Hence Germany." Another pause. "So I suppose my going to the US was almost inevitable. But first, I got my Diplom from University of Göttingen, at age 24."

"That's early." Frank acknowledges. She nods, yes.

"After that, I wanted to go to where the most interesting science was happening, so I applied to PhD programs in the US and got accepted. I had an informal understanding with a well-known lab at Princeton. That seemed like the best choice." She hesitates, but then continues.

"During my PhD work was when I got really into cell death and its roles in multicellular organisms. It's so fascinating. Did you know—I guess you must—early on, no one thought of cell death in connection with tumors? The cancer field was all about growth. Growth or no growth. That seems incredible now."

"Yes. It does, doesn't it? Fields really do change. Seen up close, science seems to move only by these very small, incremental steps. But larger changes also occur. Like realizing the role of cellular suicide. And the pathways involved."

"That's what I did for my PhD, I found CED-14, the last element of the core apoptosis pathway. I found it in worms, of course, but I also showed how it worked in human cells."

"Oh, so that was your PhD work?" Franks lights up. "Fantastic. That was such beautiful work. I remember reading the paper, amazed at how elegant and definitive it was. I almost regretted leaving active science then, wishing it had been me. So you were in Howard Wilson's lab, then?"

"No, I was in Hannah Schiller's lab, at Princeton." Chloe says, with chill creeping into her voice.

"But I thought. . . I distinctly remember reading the paper. It was a letter in Nature, wasn't it?"

"Well, yes, the Wilson lab had a Nature paper. I had the story as well but I ended up publishing in Development, some months later."

"Oh, sorry. I didn't realize there were two papers." He says, silently cursing himself.

"No, people don't. They just see the Nature paper. And they attribute the discovery of CED-14 to the Wilson lab." The bitterness in Chloe's voice is palatable. Frank remains quiet for a moment.

"But now you have a Nature paper on the way, a full-length article. So it's all good." he tries. "Your father, being a doctor, must be very proud of you."

"I guess so. But why are you so interested in my parents? I found my own way." Chloe says with obvious annoyance. She pauses, changes tone. "So, what about you? Are your parents content that you left science after your PhD? That you became a newspaper reporter instead?"

"You mean, do they mind that I am a "failed scientist"?"

"I didn't mean it that way."

"I think maybe you did. It is a common reaction to anyone leaving the academic track. Most people won't say it to your face, not directly. But fair enough, I was clumsy a moment ago." He pauses. "Well, as it happens, my parents are quite happy to see their son's name in the bylines of the New York Times. It is a big deal to them. And I am quite content with my choices."

"Of course. I get that. And serious science journalism is important. Otherwise how would the paying public know what is going on?"

"Much obliged. Anyway, the opportunity to really explain the scientific discoveries is why I worked so hard to get this particular job. As long as I can make it clear why something is exciting, our readers are curious enough to keep reading. Of course, if it looks like there is potential for development of a new cancer drug that sparks some extra interest."

"Understandable, I guess."

"Yes. A bit of 'human interest' helps as well. Everyone is curious about the scientist who found something important. So I ask about background stuff and include a nugget sometimes. A perk of the job—I get to meet lots of interesting scientists along the way." He attempts a smile. But Chloe stays serious.

"But don't you miss the lab? Don't you miss making the discoveries? It is such a rush when you realize you have found something new, something important."

"Sometimes, I suppose." He responds with a weary smile. "But then I remember the slow slog of it, as well. No, I'm better off where I am. And there is always something new and interesting to learn about."

Chloe appreciates this. Trawling through all the new discoveries and talking to the key people must be interesting, in its way. But she could never see herself doing a job like that. She needs to be the discoverer.

"Frank, thanks for the coffee" she says, starting to get up. "But I have to be getting back. Work to do."

Her mind has gradually returned to the experiments she wanted to get started today. An additional twist, another set of features to test, just occurred to her. Yes, perfect. She should get to the lab as quickly as possible.

"Thank you again for your time." Frank says "And good luck with the job interviews. I'm sure that you will take them by storm."

She hopes he is right. Her first interview is next week—and several more lined up for January. This will be a crazy, exciting time.

Chapter 6

In her fury, Karen hardly notices the biting January cold. Once buzzed in, she heads directly for the Küstner lab. She just wants to get this over with. In the microscope room, everything is as she left it last night. Even the image on the screen looks familiar, unfortunately. As she had feared, no overnight recordings have accumulated. She did indeed forget to activate the program last night after the manipulations. How could she possibly forget such a simple thing? Cursing herself, she sits down heavily. Stupid. Stupid.

After a few moments, she starts cleaning up. It's just one experiment wasted, she tells herself. It probably would not have worked anyway. Cells would have been damaged or the connections not fully gone. This cutting experiment is proving to be very difficult, indeed. After almost 2 months of trying, it is still not working reliably. It is so aggravating. Tom suggested that she should write up a paper now. All the observations and correlative evidence she has to support her ideas would be sufficient for a decent paper. It might not be enough for a top publication, but it would be publishable, for sure. The problem is, she needs a top publication. She also needs the satisfaction of doing the right experiment and getting the answer. Once she had this experiment in mind, doing anything less seems like a cop-out. She will make it, she will. She can set up another experiment tonight and it will be like this morning never happened. Perseverance is key.

Next stop is the tissue culture room. It is still empty. Karen washes her hands, takes out medium to pre-warm and wipes the hood. She has made two specialized cell lines for these experiments and both need to be in good shape to set up more trials. She takes two tissue culture flasks out of the incubator. The first one looks normal, but the color is off in the second one. The liquid is yellow, not red. She closes the incubator and tips the flasks, holding them up to the overhead light. The yellow color could mean trouble. Looking more closely, she can see that the liquid is indeed turbid, possibly contaminated. She quickly seals the flask. Oh, no, no, she thinks. This cannot be true. But it is. When she places the flask on the microscope and takes a closer look, she sees, in addition to the large sedentary cells, a dense cover of little rod-shaped objects bouncing around in the medium. Yeast. Oh, shit, she thinks. Maybe only this flask is affected; the first one looked OK by color. But here, as well, the microscope reveals the characteristic mini-rods bouncing around. With a sinking feeling, she opens the incubator and pulls out more flasks, from further back. The microscope shows her more of the same. Bouncing little rods, multiplying happily in the rich medium. The contamination is widespread. Yeast, impossible to get rid of. This cannot be happening, she thinks, not on top of everything else. She steps backwards, leans against the wall, and lets herself slide down until she is sitting on the floor. One of the offending flasks is still in her hand. She stays there, unwilling to let the day progress. This is just not fair. Another setback. Sometimes doing science is truly awful.

After a few minutes, she picks herself up and starts throwing her flasks, dishes and medium into the biological waste. It has to be done. The only way forward is to be tough. Throw it all out and start over.

Allison comes in when the carnage is almost over.

"Good morning" she starts but then notices the extra biological waste bag and the look on Karen's face. "Or not, it seems. What happened?"

"Yeast, in all my cell lines. I must have contaminated the medium or something and it has spread around. I have to throw everything out."

"But why everything? Maybe some are OK."

"It is not worth it. I tried that once. You continue with the flasks that look OK only to find out a week later that they were just less contaminated. It just takes one yeast cell. And in the meantime you manage to contaminate all the fresh reagents and cells. No, I have to do this properly. Everything out. Start with fresh cells from the liquid nitrogen storage."

"Poor you. At least you have frozen cells, so nothing is totally lost."

"No, just weeks of my life." Karen says, but catches the bitterness in her voice. "Sorry, this has just been a very bad morning. Actually, it has been a bad month."

"Is your cutting experiment still not working?"

"I am still struggling with it. And now this. Sigh. It is so depressing. Lab work is so finicky, sometimes. Why does one become a scientist, anyway?" She shakes her head. "I shouldn't be saying this to you. Sorry." She manages a weak smile in Allison's direction. "How are you?"

"OK, I guess. Alex is super-busy in his new job. Maybe it is good that I am not there to cramp his style. We talk every night, which helps. And my experiments are working fine, so I'll be done soon. Almost ready to write up my PhD thesis. But wait—should I be worried about this contamination?"

"I don't think so. I use my own reagents." Karen hopes she is right about this. If other people's experiments are affected as well, she will become very unpopular, very quickly. "Why don't you wash up and we can have a look?"

It does not take them long to establish that all of Allison's cultures are fine. No contamination. They are both relieved. Karen returns to her cleanup job, sealing up the waste to be autoclaved. She is all grim efficiency.

"Will you come to the seminar at 10?" Allison says to Karen's back. "It's this really cool stuff about synaptic memory, how experience changes the synapses in the short term and the long term. It's Henry Green, from the Salk. I'm thinking I might apply to his lab for a postdoc. Plus I've heard that he gives really great talks. So, please come, OK? So you can let me know what you think of him and his science."

Karen suspects Allison is trying to take her mind off her current disaster. She feels not the least bit motivated to go to the talk, however wonderful it will be. But she appreciates Allison's intentions and decides not to disappoint her. "Sounds interesting. I'll be there."

After the seminar, Karen is eating an early lunch alone. As Allison had promised, the seminar was excellent. She almost didn't go, indulging her black mood. But it was good that she did. Hearing some good science helped give a bit of perspective to her present practical problems. Nothing has gone badly wrong. It will just take a little more time. Visualizing the afternoon's work, she picks up her tray and heads back to the lab.

In the hallway leading to the lab, Karen sees a new item on the noticeboard. It looks like a half-page newspaper clipping, carefully placed right in the middle of the board so it is impossible to miss. As Karen gets closer, she sees the picture first. It is a "scientist at the bench" picture, complete with lab-coat and a gloved hand pointing to something on an autorad. Chloe is smiling ever so slightly and looking very confident. So, in addition to everything else, Chloe is photogenic. The picture looks like a model pretending to be a scientist. But it is the real thing. And it is in the New York Times. This must be a story about Chloe's Nature paper. And an article in today's newspaper must mean that the paper itself is now out in Nature. "Young scientist finds Achilles heel of tumor cells", the headline reads. Karen does not read the newspaper article. She knows the story. She moves on quickly.

The afternoon is busy, but not busy enough to keep Karen's thoughts from straying back to Chloe's news clipping. After this morning's setbacks, Chloe's confident smile seems to be mocking her. She feels a surge of hopelessness, a sudden certainty that it is not possible for her to get to where she needs to go. Maybe she is not cut out for this. Why is it so easy for Chloe? And why does everyone make such a big deal of her story, anyway? It is just another piece in the puzzle of how Myc is regulated, nothing fundamentally new. But she got it into Nature and now the pretty girl gets to smile to everyone from the pages of New York Times. She is probably charming them all over the job market right now. Karen takes a deep breath and shakes her head. Stop it. She needs to get a grip on herself. It's just a bad day. Just do the needful, like Bill says. She forces a smile. It almost works.

That evening, Karen gets home a bit earlier than usual. From their tiny entryway, she can see Bill in the kitchen, with his back to the door. The bag on the kitchen counter suggests he has been shopping. He turns as she enters and a warm and welcoming smile spreads across his face.

"Hi—you're home early. That's nice." He says. The oppressive silence from this morning is long gone. "I was going to cook us a nice dinner. Maybe we can have some wine, even though it is a school night."

From the quick look he gives her before turning back to the food she guesses that he wants to forget about last night's argument as much as she does.

"That sounds perfect." She says, stepping over behind him and giving him an extra-long hug. She is relieved, almost to the point of crying, to have Bill to come home to today. "I'll open the wine. I really need a glass. I had a truly terrible day."

She finds the Chardonnay in the fridge, pours them each a glass and places Bill's next to the cutting board that he has out already. Hoisting herself onto an empty section of the kitchen table, she commences the tale of the day's disappointments. Forgetting to set the microscope last night is recounted but its impact has already faded. The yeast contamination was so much worse. "I have to start all over with new cells—it sets me back at least 2 weeks. And I just feel so stupid and hopeless and I don't know what." With the last comment her voice drops off. Bill stops the food preparations and turns around. "You were just unlucky. No stupidity involved."

"But that's what I feel, stupid. With my experience, I shouldn't be making these mistakes."

"Oh, come on." He says gently. "This can happen to anyone. Don't you remember back in Paul's lab? It happened to me twice in 4 years; once just when you got there, I think. I had to restart everything too."

"Right, you were the one to show me how to thaw cells." She smiles faintly, remembering. "I had just arrived. It was in the third year of my PhD. It is embarrassing to think about, really. I was so determined to show that I was not a beginner. I probably told you this in no uncertain terms. You know, back in Nils' lab, I was the only one who had ever set up cultures of primary cells. How naïve I was, or arrogant."

"But my method for thawing cells was better."

"Your method was better. I remember. I used it today."

"Good old Bill, at least he can teach the visiting students how to thaw cells." Bill says, his self-depreciation not completely devoid of bitterness. But Karen follows the recollection in a different direction.

"I remember coming to Boston, to this lab full of smart and self-assured scientists. The lab itself looked pretty much like my first lab. In fact, the benches were older and equipment was more outdated in Paul's lab than at Karolinska. Yet it was a shock, this brave new world. Everyone was so smart, determined, hard-working. It was overwhelming. But it was also exciting. I suddenly felt a lot more alive." She drifts off.

"But all those bright young things also got yeast in their cells. Not just me; Akira, Peter and Catherine, who all went on to do really well. So you are in good company."

Karen remains silent for a while.

"I still can't help thinking that it's because I haven't been paying enough attention. These last couple of months, all kinds of things have not been working. Maybe I'm losing it."

"Come on, Karen. You are not losing it. This is just a rough patch. You've been there before. Don't get so discouraged."

"I just seem to spend all my time starting over. Today it was just too much." She draws a heavy breath. "But you would have been proud of me. I didn't break down or anything. I did what I had to do. I threw everything out, I cleaned, I got started again."

"Don't you think—" Bill starts, in a gentle voice "that all these things going wrong could be related to you driving yourself so hard?" He gets no immediate answer. "You didn't even take time off for Christmas this year, not really."

"I had to.." She starts.

"I know," Bill continues "the microscope being available and all that. But everyone eventually tires if they keep pushing too hard. Maybe you could take it easy for a couple of weeks, while the new cells are getting going. And we could go somewhere for the weekend, perhaps."

Karen knows that she should give in to Bill's spousal ministrations. She knows that he is right.

"Yes. Let's do something nice this weekend. And I will stop moping about my lost cells. As of now." She slides off the table. "So, can I help with a bit of chopping here?"

After dinner, Karen's thoughts start drifting back to the depressing day at the lab. There was one part that she did not tell Bill about earlier.

"Did you see the article about Chloe—in the New York Times?"

"No, I haven't seen the paper today. What was it about?"

"Clever Chloe takes the world by storm, smiles charmingly—and cures cancer. It seems her article in Nature came out this week."

"So is that what's making you extra grumpy today?"

"Maybe. A bit."

"You know that this has nothing whatsoever to do with you."

"I know, I know. It just makes me feel extra hopeless. Chloe gets Nature articles. I grow yeast."

"Maybe you should be a baker." He smiles, weakly. "Come on, Karen, I can see why you would be upset if you were working on the same thing as Chloe. But your work is completely different." Karen nods, yes. "And much more original." He adds with a smile.

"Bill, I really appreciate you being so supportive. I do. But the fact is, she is interviewing for jobs and I am still running around in circles."

"You are not at the same stage yet. Your time will come."

"I just find that really hard to believe right now. Maybe I just have to accept that there are people who have all the luck and all the charm and will get all the great jobs. She even looks pretty, the poster girl for science."

"That's just silly. Who cares what she looks like? And newspaper stories are not important in this business. You know that." Bill protests.

Karen continues as if she hasn't heard. "Other people, ordinary people, are second-rate. We get the little papers and some stupid job at some nowhere place and no one will care what we do."

"Now you really are being unreasonable. You are not ordinary, not second rate. And Chloe having a top paper is neither here nor there."

"Right. Yes, of course, I know." Karen says, but with no sign of being convinced or uplifted. After a few moments, she gets up and starts to clear the table. Bill has been more than understanding. He should not be taxed further with her stupid insecurities tonight.

The next morning, Karen is the first person in the lab as usual. Not a soul to be seen anywhere. Bill tried to suggest a slower morning might be good for her. But she knows she needs to keep doing what she can—at least until she gets on top of everything again. Bill let it be, after extracting a firm promise that the weekend would be different.

Switching the overhead lights of the lab on, all the desks and benches spring to attention. Most mornings, Karen does not notice the other benches on the way to her own. She is too preoccupied with the day ahead and the experiments to be done. This morning, however, she stops at the second bay. It is Juan's and Chloe's area. Karen is drawn to the tidied bench of success. Part of this bench was visible in the picture showing Chloe "working". Karen slides two fingers along the edge. It is at rest now. Chloe is on her long interview tour. Karen looks at the clear bottles on the shelves above the bench. 10 mM Tris pH 7.4, PBS, 10 % SDS, the same things they all have at their benches. There are no secrets here, of course not, just lots of ordinary bottles. The drawer under the bench is a jumble of special markers, colored labels and manuals, exactly what you would expect to find. The small fridge further under the bench contains several stacks of slide holders, all full it seems, and underneath them, bacterial plates and racks of colorful tubes. She closes the fridge door gently and slides further into the bay. Surely the desk is the key to the work and the success. A neat stack of printed papers sits to one side. She gently lifts up one, then one more. She recognizes a recent Cell paper from another lab and a recent review by Tom and a former postdoc. So, she is keeping up with the lore of the master. Both look well read, as do the papers below. The corners of the pages bend upwards and the text is decorated with underlines and scribbled words. Also on the desk are 8–10 gray lab notebooks, arranged neatly with the blank spines facing out. A monthly calendar pinned on the board above catches her eye. January is proudly displayed. She sees "UCSF" and "Stanford" written in red and blue pen, respectively. There are several additional, harder-to-read markings. She leans in closer to read.

"Are you looking for something?"

Karen pulls back quickly, in reflex. Juan has come in and is standing at the other end of the bay. He must be on his way to his own desk and is obviously surprised to see Karen at Chloe's.

"I was just, just wondering. . . . I was just wondering when Chloe will be back"—a tinge of panic in her voice. Her desperation is probably nakedly obvious. "I was going to ask her about something."

Juan looks at her, clearly suspicious, but also uncomfortable. "Chloe left just 2 days ago. She won't be back for another 3 weeks." He finally says. He does not move any closer, just stands there, looking. For a few long moments, they both seem paralyzed.

Karen looks down, mumbles "thanks", and slides quickly along Chloe's bench, past Juan and out of the enclosure that the bay has become. Without looking back she hurries to her own desk and sinks into the chair, out of view. The panicky, tingling feeling threatens to erupt into visible shaking or worse. She puts both arms on the desk to steady herself, and lets her bag slide into her lap.

She stares at her desk for a long time without seeing. Will Juan tell Tom? Or Chloe? Or anyone? One does not go snooping around at other people's desks. It is simply not done. But when did Juan come in? Maybe he is not sure what he saw. She vows to avoid him for the rest of the day, for longer if possible. What on earth was she thinking? Crazy. Stupid. She pretends business at her desk until she feels sure that Juan is staying put and she can leave the main lab unseen. Out of habit, she hurries to the tissue culture room. Juan is not using it these days. She will stay here for a while, she thinks, clean the hood, check her cells, anything to stay busy until she feels confident that she can return to the lab and act normal. Then she will get on with the day, somehow.

Chapter 7

All three hoods are busy. Tissue culture flasks, tubes, bottles and pipettes are being skillfully manipulated by industrious hands. The noise in the small room is deafening. With conversation almost impossible, each hood becomes a small bubble of privacy, allowing long, uninterrupted trains of thought.

Allison rolls her chair back and looks around, in need of distraction.

"Karen. How are you? That was a real bummer, wasn't it?"

"Allison. Hi. Fine, thanks." Karen was far away, but manages to suppress her irritation at being disturbed. She wonders what Allison is referring to, and then remembers that last time they were in the tissue culture room together, 2 weeks ago, she was throwing out contaminated cultures and reagents.

"I've dealt with worse before." Karen continues. "That's just the way it is. You have to get back on the horse, you know? I'm ready to go ahead with my favorite experiment again. Soon."

"So back to normal overdrive, then?"

"I suppose so," says Karen. "I just need to get this experiment done, so I can write up my paper." She has enjoyed the short respite from the punishing hours. Bill was right. She needed to take a step back and refocus. This may be why she feels so ready to tackle the tough experiment again. "I just have to make it work this time. I think I know how. I talked to Greg, you know, the microscope guy in Bettina's lab, and he has some good ideas."

They both work on in silence for a while.

"Lunch later?" Karen finally suggests.

"Sure. The usual—a quarter to twelve?"

Walking back through the main lab an hour later, Karen looks straight ahead and locks onto her bench. She does not look down Juan's and Chloe's bay, as she has forced herself not to do every day for the past 2 weeks. But she feels a sharp twinge of shame as she passes, anyway. Snooping around was a moment's craziness. She was so upset, unbalanced back then. She is better now. Luckily, it has been easy to avoid Juan. And Chloe is still not back from California. Karen would prefer for her to stay away, for her to get on with her successful life elsewhere. Seeing Chloe's cheery face again, hearing her confident, eager voice, will only make it harder to forget that strange morning.

"I'm not going to work on cats or monkeys or anything like that." Allison explains "I want to keep working at the molecular and cellular level. There are some pretty cool cellular models, like the simple reflexes in *Aplysia* neurons. It is incredible." Alison's eyes widen "Or maybe I'll work with *C. elegans*. With all the genetic tools and now the possibility of switching neurons on with light pulses, you know, optogenetics. They don't look like much, but they have some neat behaviors, like when they move?" She wiggles her hand in a forward direction and lets it avoid an imaginary obstacle. "It would be so cool to map out a circuit and try to find out how it changes with learning and memory."

Over lunch, they are discussing Allison's plans for her postdoc.

"Sounds really interesting. Have you written to any of the labs?"

"Not yet. I wanted to wait until my major paper is accepted."

"Well, don't wait too late. It takes time to get into a popular lab."

"But I need a good paper first, don't I?"

"You could always ask Tom to write to people and explain that it's coming. They will trust his judgment and at least interview you."

"But you can't be sure until the paper is accepted, in press, can you? If you get scooped, it will be so much harder to get a fellowship." Allison's eagerness is gone. She seems to have deflated.

"Why are you so worried about getting scooped all of a sudden? Of course, you can never know for sure whether someone else is working on exactly the same thing you are. But it is rare."

Allison shifts in her seat and looks increasingly distressed.

"Promise that you won't get mad at me?" She says, in a quiet voice.

"Mad at you, why should I get mad at you?"

"I should have told you earlier. But I didn't want to be the messenger."

"Told me what?" Karen straightens in her chair.

"There was a paper in last weeks issue of Science, from a Polish group. I thought you knew."

"Knew what? Which paper?"

"A paper on nanotubes and their role in cell survival." Allison says, enunciating very clearly. She keeps her eyes on her hands.

A sharp intake of air. Karen stares at Allison, then shakes her head, blinks and looks off into space, not responding.

"Shit" she finally manages "so I have been scooped, have I?" She sits completely still. "I don't even know anyone working in Poland."

Allison does not say anything.

"I guess I should go have a look." Karen says, with forced calm, as she gets up with her lunch tray. She is stiff with the effort and continues to gaze into the distance. She does not look at the tray, or at Allison. An open bottle tips over and drenches the table, water spilling over the side. Allison busies herself cleaning up, relieved to have something to do.

Back at her desk, Karen snaps awake her computer from its chirpy word-of-the-day display. The last thing she needs right now is to learn another fancy word. She goes to the Science web site. The paper is right there: "Ultra-thin nanotubes connect distant cells and regulate cell survival." She hesitates, willing it to go away, to wake from the bad dream, anything. But no. It is still there. She sits. One click and the article appears on her screen, complete with figures and professional formatting. This is the real thing, accepted, published and unavoidable. And obviously, it is not hers. She reads through the short article quickly and glances at the figures. It does not take her long to absorb the key facts and findings. This is bad, very bad. She picks up the phone on her desk and punches in Bill's number. Please be in, she pleads, please.

"Biology department, Bill Smith speaking".

"It's me." She says, almost in a whisper "Can you talk?"

"Hi sweetie, what's up?" He adds "Sure, I've got the office to myself."

"I just need to talk. I've been scooped. My project. It's terrible."

"Wait a minute, scooped? Where? But there wasn't anyone else working on this, was there?"

"An article in Science, last week. I only saw it because Allison told me. It is from a group that I never heard of. From Poland of all places."

"But what have they published? Exactly like your story?"

"They use different cells and a different assay. But the basic findings are the same. They identify the thin connections. They call them nanotubes. And they have some evidence, rather poor evidence, actually, that having the connections matters for cell survival. The correlations are there, that's OK. But the rest is crappy, half-assed evidence." She is suddenly furious. "They do these crude genetic manipulations and find that the connections, the nanotubes, are affected. And cell survival is affected as well. That's it. They conclude the nanotubes are required for survival. It is just a dual effect, no clear evidence that the connections, as such, actually matter. And they got a Science paper out it. I can't believe it."

"You were so sure that you would need to go further for a good publication. I remember you telling me that over and over."

"Yeah, I was. Obviously, I was horribly wrong. But I wanted to go for a top paper. And now this group from nowhere manages to publish with such crappy evidence. I just can't believe it."

"Poor you, sweetie. I feel terrible for you. But what you have is so much better. Surely you can still get it published? Especially with the cool experiments that you're doing now."

"I am still not done, remember?" Her voice is cracking, turning shrill. "I'm still trying. It's so pathetic. And even with the most fantastic evidence, all I could get now is a 'me-too' story, in some crummy journal. Once something is published in a top journal is it considered 'known'—even if they did everything fast and sloppy. You know that." She is aware of the misdirected anger in her voice, but unable to stop herself. "It will probably take me a year just to get a story out, even in a minor journal. And if anyone actually reads it, they will think I got the idea from the Science paper. No one will ever know that these were my ideas, my observations. It will look just like any other sad piece of derivative work. And that's the one thing I swore I wouldn't do. Derivative work. Sometimes I just hate this world."

She stops up. "I'm sorry. You don't deserve this. I just feel like giving up the whole damn thing, I really do. I should talk to Tom."

"Why don't you wait? Wait until tomorrow, when you've had time to think." Bill suggests, sensibly. "Pack yourself up and go home. I'll meet you there in an hour or so. You can talk to Tom tomorrow. Don't do anything dramatic, OK?" He is the voice of reason and so immensely considerate. This is why she loves him so. This is why she called him. She lets herself be persuaded.

Karen stands up and gets ready to leave, not making eye contact with anyone. They may have seen the Science paper and remember her project well enough to understand her predicament. She does not want their sympathy, or pity. They will be relieved it is not their work that has been scooped. Naturally. There is no ill will from them or from her. She just prefers not to talk to any of them. It won't help.

Hit by the early afternoon sun, the large glass doors of the institute are like barriers of light. Karen pushes through, with more force than really needed, and encounters a cold and bright winter afternoon. It seems like ages ago that she came in for work this morning. She was finally feeling a bit better. Now everything has been turned upside down again. She is beaten, shattered. Her big story, snatched away. Two years of work, wasted. Her long battle with the perfect experiment, wasted. It's unbelievable. How incredibly lucky they were to get their story published in Science, she thinks, with such shaky evidence. The reviewers must have been sleeping. But it makes no difference. She cannot change it.

Without noticing where she was going, she has followed her normal route to the bus stop. Surprisingly, the bus turns up almost immediately. She climbs aboard and finds a seat in the back. She steers her mind ahead, to the afternoon and evening, cocooned in the comfort of Bill's support. What would she do without it? Who knows where she would land from this free fall. In a deep depression, probably. She would have to leave the country when her fellowship runs out, and go back as a complete failure. As it is, she might be a failure, but at least she has Bill. No, she will not let herself be a failure, she thinks with sudden force. She will have to find a way, even if it means starting all over on a new project. She will. She will. The surge of determination dissolves almost as quickly as it appeared. Her thoughts return to the Science paper. It sucks the air from her. She thinks of her battle with the cutting experiments. How could she be so stupid and not try to publish earlier? Tom did suggest that. And why could she not make the damn experiment work? Pathetic, it's pathetic. Who does she think she is kidding? Bill says she is great, but he is not exactly objective, is he? He doesn't know. Just look at the other postdocs in Tom's lab, the ones with the perfect backgrounds, always making the right moves. Like fucking Chloe. I'm such an amateur, Karen thinks. She slumps back in her seat. It might be all up from here, from the bottom of this awful pit. But it is a steep ascent. And she is not at all sure she is enough of a mountaineer to brave it.

Two days later, Karen's thoughts are still in turmoil. One moment she is ready to throw it all overboard and leave the project, leave the lab, leave science forever. The next, she is calm and sensible, considering what can be salvaged from her project. Tom has been away, so no help there. To make matters worse, she is scheduled to give group meeting today. She does not want to get up in front of the whole lab and talk about her lack of progress. Maybe everyone will feel sorry for her, because of the Science paper. Even worse. She sent an email to Tom to ask if her group meeting could be postponed. But he wouldn't hear of it. He said that he would be back today and that a group brainstorming might be just the thing she needed. She should just go ahead. He obviously does not know what it is like to get up in front of everyone and not have enough to say. But she does not have a choice now. It is almost 2 P.M.

The group meeting goes on for a long time. Karen has explained her results so far and starts on a short review of the Science paper as a basis for a group discussion about possible ways forward. This was Tom's idea, communicated in his email yesterday. She is still not convinced it is a good idea. It gives the impression that she does not have ideas of her own but has to rely on input from others to move on. It is embarrassing, really. And what if there are no suggestions? Or even worse, if she gets only useless suggestions, time-wasters? Can she ignore them, without offending anyone? But Tom is the one with experience. And she desperately needs to move on. So she is doing what he suggested. She tries to present the Science paper in an objective way, explaining what they have shown but also some of the shortcomings, including the lack of an experiment that directly tests the proposed role of the nanotubes.

Surprisingly, Tom takes the word before she has fully finished. He seems impatient, either with her or with the so-far silent group, or both.

"Look, Karen. I agree that it would have been a better paper if they had addressed these issues. But that gets us nowhere. Our question today is how you get a publishable paper from what you have. Even if you solve your problems with the cutting experiment, this, by itself, may not make a story that can be published well, not at this point." He pauses, briefly. "We should consider all your data and determine what your unique angle might be. And then see if you can boost that angle into a set of coherent results that can carry a paper. If you have that, you can get your novelty-compromised data published as part of the story. So, what do you have that is not in their paper?"

"Well, I suppose, the difference between normal cells and cancer cells in my assay. But I did not follow this direction very far. The thin connections, the 'nanotubes' or whatever, are clearly altered in transformed cells, both in number and dynamics. But I don't know if this difference is meaningful."

"And cell survival?"

"Also different, of course, between normal and transformed cells. It generally is. But I didn't use my live sensors in those experiments, so I can't be sure how survival relates to cell connections in transformed cells. But I could of course..." Karen offers, tentatively.

"And in the Science paper, do they discuss this?"

"No, well, yes, they suggest that nanotubes might be different in transformed cells, but they have no data. It's just speculation."

"Speculation is cheap," says Tom, more keen now. "Showing is a whole different thing. So how could you wrap up some nice data on this, quickly? Apart from putting your sensors in transformed cells and running the 3D assay, which I agree is a good idea. Any suggestions from the crowd?" He looks intently around the table. The direct challenge seems to rouse people and some suggestions emerge. One is that it would help to get some data from mouse tumor models. Several of them have noticed that the Science paper contains no tissue or animal data, only cell culture. But making transgenic mice or getting mice from others takes a long time. Karen needs something fast in order to submit a paper soon.

In the end, it is Vikram who comes up with the best suggestion:

"Why don't you try the MMTV-Ras plus Myc mice that Chloe used for her paper? I believe you simply cross MMTV-Ras and MMTV-Myc carriers and look at tumors in the double transgenic offspring. It's a reliable tumor model that has been around for years. It was developed in this lab, wasn't it, Tom?" Tom nods but does not interrupt. "The tumors develop quickly. And it's in the mammary gland, so you might be able to image the tissue with a skin flap.

We have the set-up in the facility, I think. You could look at the normal tissue in the same way."

"I could try that. I have done tissue imaging using skin flaps before." Karen narrows her eyes in concentration. "It may be doable. My markers for cell connections are very bright and I have them in viral vectors, for short-term transduction. Cell death, I can detect in several ways. I would just need to recognize the tumor cells in the tissue."

"Oh, that's easy." Vikram waves a hand dismissively.

"Could you show me how, to start out?" Karen says, looking directly at Vikram. She is surprised at herself, asking for help. He shrugs, sure. Karen looks at Tom. "I would need some double transgenic mice, though."

"Not a problem." Tom says. "We should have breeding populations of both strains. The reviewers of Chloe's paper asked for this experiment, so we worked with the mice not very long ago. Chloe is not back from her tour yet, but Andy should have the mice. Go ask him, Karen. Tell him that I said to give you everything he can spare. That was an excellent idea, Vik, excellent."

"Thank you" Vikram tilts his head in a hint of a bow. "We aim to please."

"Right. Yes. Thank you." says Karen.

The meeting wraps up soon after that. Karen hurries to make sure she can catch Andy before he leaves for the day. On the way out, she asks Tom. "Are we still on for talking tomorrow morning?"

"Absolutely. My office at around 9?"

When Karen gets to the mouse facility in the basement, Andy is not in his small office. He must be inside. Feeling the need for action today, she decides to suit up to enter the facility. In the small changing room, she exchanges her shoes for a pair of clean, shared clogs and puts on a disposable blue clean-suit over her clothes, with a hood covering her hair. Two rounds of hand-washing, a pair of blue gloves to be safe and she is ready to go in. It seems over-protective when all she needs to do is to talk to Andy. But these are the procedures and she does not want to rub anyone the wrong way. Not now, when she needs help.

She finds Andy in the second of the mouse rooms. He does not have any animals out, so she figures it is OK to talk to him.

"Andy, Hi, sorry to interrupt." Karen starts. He turns around with a puzzled look. She realizes that she is partially disguised in the blue suit. "It's Karen Larsson, from Tom's lab."

"Right, what can I do you for?" He says.

"I need some MMTV-Ras, MMTV-Myc double transgenic mice and Tom said you might have what I need. I suppose you keep them as heterozygous single-transgenic lines? So I need carrier males and females of each that can be mated." Andy looks at her, nods. He must be waiting for more. "Tom thought you would still have both strains going. They are the ones that Chloe used recently. Tom said I could have access to whatever you can spare. I'd really like to set up some matings as soon as possible, maybe 6 or so?" She pauses, considering numbers further. "Depending on litter size? I'd like to have 10 or 12 double transgenic pups."

"Right. Well, the double transgenics are hard to get, so for that many pups, you need to order in advance. You let me know 3 months ahead of time. I expand, do the matings, and another 3 weeks you get the pups for your experiments. Like I told Chloe back then. That's the procedure."

"But she did it quickly, Tom said. I can't wait the extra three months. It's just as important to me as it was for Chloe." Karen realizes she might sound a bit whiney. But she cannot wait half a year for a result.

"Well, normally, that's the procedure. But I think you may be in luck. Let's have a look." Andy says and walks toward the computer in the hallway. Karen hurries after him.

"Let's see. Chloe got what I could give her back then. I only had two MMTV-Myc females to mate with MMTV-Ras males. I seem to remember one of the pregnant females died, though, before term. Anyway, I expanded both strains, so you are in luck." He shrugs. "Chloe was really pissed off that I didn't have more mice available for her, so I assumed she would be back for more. I haven't seen her for a while, though. If Tom says it's OK, then you can have whatever is available."

"Thank you so much. I really appreciate it. How many can breed soon?"

"I have plenty. You would need about ten mature MMTV-Myc females to be reasonably sure to get ten double transgenic offspring. I have that and more coming along. It is best to cross them that way. Even with the Myc mothers, the litters are small, with Ras it is much worse. As for double transgenics, well, you tend to get less than one in four live pups."

"That sounds perfect, then. Should I help you with setting up?"

"I can take care of the mating and keep an eye on the pregnant females until they have pups. But you do the genotyping. That's the arrangement I have with Tom."

"Of course. Thanks so much. This is really a big help."

"Just remember to put the request through the system immediately so there is no problem this time. And get Tom to sign off."

"Sure, of course."

As Karen is changing back out of the clean suit, she remembers something Andy said. Two females, and one died. How could that possibly give enough double transgenic offspring for a decent experiment? Something doesn't add up.

Back at her desk, she opens the pdf of Chloe's paper. It is in the last panel of Figure 7, Figure 7F. It is a simple experiment, three MMTV-Myc, MMTV-

Ras double transgenics with drug and three without. The effect of the drug is clear-cut, so three of each is enough. But this means six double transgenic animals. How could she possibly get six from one surviving female? Even with a decent litter size, this is essentially impossible.

Karen stares at the graph displayed prominently on her screen. It's not possible, she thinks. Chloe must have cut a corner on this very last experiment. If she did, if she fudged the numbers, then this is very serious. But, be careful now. Maybe Andy is wrong. Maybe he had more mice than he remembers. But he was looking at the records when he told her. Maybe both females survived. But even with two females, six double transgenics is very unlikely, especially if the litters are small. Anyway, remembering that one female died seems so specific. Why would he say something like that if it were not true? He doesn't know what the mice are used for and how the papers end up. That's just us, she thinks, Tom and us. She will have to tell Tom about this. He needs to know. Or what? Maybe she should keep out of it. Andy was not definite about one female dying, was he? And with two females, you could just possibly get six double transgenic pups. But it is still so very unlikely. She could calculate the probabilities. But that does not really help. Either Chloe was incredibly lucky, like winning the lottery. Or she cheated.

Chapter 8

Chloe turns to face the audience. It is a large auditorium and yet it is completely full. Latecomers have already started filling the upper aisles and populating the steps. It is amazing, she thinks, that so many people have come to hear her talk, talk about her work. All these eager faces. What a thrill. There is a camera set up in the middle aisle, as well, pointing straight at her. She has been warned that they would be recording the talk, "For anyone who cannot be present today." Or for rewinding and closer scrutiny, she thinks. Once she imagines the eye of the camera as just another member of the audience, she stops worrying about it. The introduction is done and the audience is quiet. She is ready.

A deep breath and she starts: "It is often said that cancer is not one disease, but many. Given this, one might expect the 'holy grail' of cancer research, finding a cellular activity that is required for tumor survival, but not for normal tissue maintenance, to be an unobtainable goal, a myth. If, on top of this, we require it to be a drug target, we are being very ambitious. Today I will tell you of my recent efforts to reach this goal."

It may be an over-dramatic start, Chloe knows, but it works. It makes the audience extra curious about what is to come. Her paper has been out for

3 weeks now, so they may already know the story. But, surprisingly, this seems to make people more interested in hearing her talk, not less.

The well-rehearsed words continue, almost without her conscious intervention. She makes loose eye contact with the first few rows of the audience. It engages them and the visual feedback allows her to fine-tune the delivery. This is miles better than her first job talk, more than a month ago. Since then, the slides have been carefully trimmed and perfected. The talk has been rehearsed out loud many times, most recently this morning in her hotel room. It has been slightly modified after each of the formal deliveries in the past month, tweaking it to remove bumps in the flow. It surprises her that the many rehearsals and deliveries do not make the talk worn or stiff. Nor does the talk come out exactly the same way each time. Quite the contrary, the certainty of control, of knowing what is to come, liberates her from any need of a script. She can focus on an engaging delivery of the story and the facts. And it works. The audience is captivated. They applaud enthusiastically when she reaches the end.

Hands go up immediately—an excellent sign. She starts by taking questions from the front two rows. They are all faculty members, it looks like, and some of them will be members of the search committee. There are some very bright scientists among them, so she has a chance of getting some excellent questions. Good questions, whether probing, suggestive or open-ended, can be goldmines if she is able to use them wisely. She needs to be on her toes.

"You show quite convincingly how important Jmjd10 and its modulation of Myc activity is for growth and survival of tumors. But surely, this activity must have evolved for a reason beneficial to the host, to us. So, I have two questions: One is what the normal role of Jmjd10 is for development or physiology. The second is related, in a way, assuming your drug is specific, would inhibition of Jmjd10 not be harmful in some way? Oh, and finally, sorry, one more question: How about escape? Do tumor cells become spontaneously independent of Jmjd10 as seen for so many other potential drug targets?"

These are all questions Chloe is happy to get. Answering them will allow her to go beyond the content of the talk and discuss the future plans that she is most excited about. It will also allow her to describe some of her unpublished, preliminary data. A perfect start. She sends a mental thanks to that reporter, Frank. Explaining her ideas for future directions to him got her started on the key experiments early. So she has some relevant preliminary data already. The first-row originator of the compound question nods along with her answers, with an appreciative smile. He seems familiar. She should find out who he is. Hands go up again. She moves on to the next question.

After many questions and answers, her host, Matt Rosen, calls an end to it. There are still several hands up from the audience, waving for attention. "Sorry, I hate to put an end to such a lively session." Matt says "But we must move on with today's program. For the students that have signed up, I remind you that the lunch meeting with our speaker is on the third floor and starts in a few minutes. You can ask many more questions there. In closing, I would like to thank our speaker, Chloe Varga, for this extremely engaging and interesting talk." There is another round of energetic applause.

Chloe is elated. Not only did the talk go well, but the question session also. For a job interview, she knows that this may be just as important as a well-executed talk. She has shown her ability to think, and think creatively, on her feet. Keeping the talk short to leave plenty of time for questions has been a good strategy. Of course, it helps to have an intelligent audience to provide interesting questions. So much better than the tepid interest at the medical school she visited in New York. She feels a sudden certainty of being in the perfect place, right now.

As people are leaving the room, there is much chatter and energy. There is a buzz of something significant having been communicated, along with an undeniable excitement about what is to come. A good Q&A session enlivens everyone, Chloe has noticed. Maybe it is the didactic nature of it, adding a new dimension to the talk. Maybe it is the thrill of witnessing a sparring match of minds. She is not sure. But she can feel the energy. A number of people head toward the front of the auditorium, directly toward Chloe. Most of them just deliver a short "great talk" or "fantastic stuff" and briefly introduce themselves before leaving. The spontaneity of these compliments feels genuine. She allows herself to enjoy the soothing rays of appreciation, before focusing her attention on the remaining people. They have more questions and speculations or specific inquiries about details of the work. She addresses one after the other, fired up and generous with her ideas.

She could go on forever, she feels, but Matt Rosen is hovering with obvious intent. He steps in again to move Chloe's day forward. He has in the meantime packed up her computer-bag and offers to take it to his office for her.

"That was a very impressive talk." He says, on the way out. "And standing room only. You appear to have a reputation that precedes you."

While they wait for the elevator Matt starts telling her about the PhD students she is about to meet. Sessions with students and postdocs were also scheduled right after the talk at the last two places she visited. It suits her well. It provides an interlude of stress-free conversation between the demanding public performance and the important one-on-one conversations coming up later. Asking them about their projects should allow her to get a bite of lunch in, as well. She needs some fuel for the long day ahead.

Celia, one of the most outgoing students at lunch, offers to walk Chloe to her next appointment. The lunch was lively and the students, with one or two exceptions, were confident and articulate. Being a self-selected group from one of the most popular PhD programs in the country, this was hardly surprising. But it pleases Chloe, nonetheless, to have them live up to expectations. They also brought up the apparently general concerns of PhD students everywhere: how to make a career in science—and whether it is worth all the hard work. This is a preoccupation Chloe does not quite understand. For her, it was always obvious. Of course she wanted the best and of course she needed to work hard for it. But she is feeling charitable today and recognizes that not all these youngsters share her level of ambition. So she simply offered up what bits of common sense advice she could muster.

The meeting ahead is a completely different thing. She will be seeing Peter Conner, someone whose work she has admired for a long time and has met once before. It is a good time of day for that: She does not have to worry about her talk and she is not yet too tired. She should be able to fire up those neurons and be alert and quick.

Gratifyingly, Peter remembers her from the meeting they both attended.

"Chloe? It is great to see you again. And that was a fantastic talk."

"Thanks. It's good to be here."

"So, I want to talk to you about your ideas for the regulation of Jmjd10 in normal cells and its role there. The stuff you mentioned after your talk." That he skips the small talk and starts on the science immediately pleases Chloe. It must mean that her work has truly engaged him. She starts laying out the details of her ideas and plans with unfeigned enthusiasm. He asks more questions and comes with suggestions, plunging into her little universe with curiosity and cleverness. Feeling his keen focus pushes her further and the conversation races ahead. They cover a lot of ground. They find some time to discuss his science as well. He talks about the new area his lab is now focusing on and shows some recent results. Occasional abrupt changes of direction seem to be his preferred approach to science. He has made several such shifts in his career and each time has contributed significantly to the new area. This is a very different approach from Tom's, she realizes. Tom's lab has always worked on the same questions, as far as she knows. The field moves ahead, of course, and the technology changes, but the focus stays the same. Both he and Peter are successful, so apparently both strategies work. She finds the new stuff from Peter's lab interesting and is alert, asking questions and volunteering ideas of her own. Although it is not her field, she knows the area well enough that she can contribute comfortably. It feels like a discussion between equals, she notes with satisfaction. They are deep into considerations of how adaptation works in his system, when there is a knock on the door. It is time for her next appointment, already. Forty-five minutes can go by very fast.

Catherine (Cathy, please) Anton, picks Chloe up from Peter's office. She is shorter than Chloe, 5–10 years older, with curly hair and intelligent eyes in a round face. Not someone she has met before, Chloe thinks. She has a friendly but brisk manner. Chloe guesses that she has children and has acquired the necessary efficiency to deal with both work and home. After an initial "I really enjoyed your talk, it was excellent," she thoughtfully asks Chloe whether she needs to use the restroom and whether she would like a coffee.

When Chloe returns to Cathy's office, the much-needed coffee is ready for her. Cathy has also positioned her computer screen so that they can both see it during their discussion. But first, she asks Chloe how she likes California.

"So far, I like it." Chloe says with a smile "But then I have only seen a few university campuses, so I am not sure I really know."

Cathy laughs and admits that she got caught in the 'bay area vortex' some time ago. "I love living here and the science is fantastic." she says with emphasis "And my husband has a good job here as well, so that sort of seals it." She shifts gears quickly. "Well, no time for small talk. Let me tell you about what we do in the lab".

Once they are into the science, Chloe realizes that she knows the work from Cathy's lab even if she did not recognize her name. Cathy has had several high profile publications recently about chromatin modifications and global gene expression changes. This is not far from Chloe's present work and interests, which makes it easy for her to provide appropriate observations and questions along the way. Cathy seems to enjoy being interrupted and enthusiastically supplies more information, thoughts and perspective in response. They get deeper into the projects and also touch on some more general issues that the field is struggling with. It is a very satisfying discussion and they could probably continue for a long time. But at one point, Cathy stops herself and checks the timer on the computer screen.

"Just as I thought. We only have a couple more minutes to talk, I'm afraid. I just want to say a few things before you are pulled away. A bit of a sales pitch, you could say. This is a great university and a great department. I honestly think it's the best. Even if I discount the bay area thing, there is no other place that I would rather work. The students are fantastic and they flock to the new labs. So you would have no problem getting your lab going here. I started about 6 years ago, and I was promoted last year. Everyone has been really supportive." Cathy adds a few more details about the department. Chloe is very happy to hear all this, smiles and nods. It is not that she had any doubts about the place. But Cathy's eagerness to supply a 'sales pitch' makes her feel truly wanted. It is flattering and very gratifying, Chloe acknowledges to herself.

As expected, they have to stop a few minutes later when Chloe's next appointment arrives to pick her up.

The next discussion is very different. Professor Meiner is in his 60s, Chloe guesses. He says "Hans Meiner, call me Hans" as he proffers his hand. But she gets the impression that 'Professor Meiner' would be more appropriate. In part it is his age, but also an old worldliness that Chloe recognizes. After the initial greeting, he turns to lead Chloe to his office. He shuffles along the corridor in a way that suggests waning energies. But this impression is put to shame once he starts talking about his science. He has worked on membrane channels for a long time, through many ups and downs of the field. It is now quite a hot topic again. Chloe finds she has to struggle to keep up. Not knowing the field in detail, an in depth discussion is more difficult than it was with Cathy. She needs every fiber of concentration that she can muster to pay full attention to his words while digging into old knowledge from somewhere in the back of her mind. It should be a conversation, she thinks, not a monologue. She asks about conformational changes and how to identify possible intermediary stages; she finds a few other opportunities for questions. He considers her questions carefully, answers thoughtfully. When he switches topic to ask about her work, she can retreat to more familiar territory. But she stays alert. She tries to give answers that open to new perspectives when possible and focuses on mechanisms rather that cancer-relevance. This may help him appreciate her work and her thinking. His questions may be more polite than penetrating but at least he seems to be taking her seriously. This is a good sign, she thinks, considering how established he is. She briefly wonders whether he might prefer for the department hire another structural biologist and not someone like her. But she cannot let herself worry about that. She can only try to make the right impression. She focuses on this for the rest of their time. It is demanding. But she passes, she thinks. Probably. Maybe.

Leaving Professor Meiner's office, she follows a balding guy, middle-aged or a bit past that. He must be her next host, Toby Ackman, although he did not introduce himself. She knows his seminal work from her student days. They appear to be going to the basement.

"Some of us are not so popular with the powers that be." He comments with a grunt and the faintest trace of a smile. "They keep us in the basement. So that we know our place."

The cozy disarray of Toby's office and his wry comments are at odds with what Chloe remembers of his papers. The dense papers are stringent and formal, a bit old-fashioned. Toby seems to enjoy complaining and sending jabs to the 'powers that be'. The PhD students and their preferences and abilities get a lashing. The promotion committee, which he says is driven by politics and grant money and not by 'thoughtful appreciation' of scientific accomplishment, also gets a turn. This makes for an awkward conversation. As an intelligent person, Toby must realize that Chloe cannot respond to any of this. She finds herself wondering if he is doing this deliberately, to see how she will react. Or maybe as the bad boy in the basement, he simply does not care what he says. In any case, she feels ill at ease. She is also getting tired and losing concentration briefly. It is like slipping in and out of dark spaces, with seconds missing here and there. It is a bit scary. She fights to keep the fatigue at bay.

Forty-five minutes later she is relieved to be escorted back out of the basement. This will be her last interview of the now very long afternoon. Her last host, 'just Barry' as he introduced himself, must be a new recruit. He is about her age and full of energy. He is tall, with brown hair, a quite handsome face and intense, but playful eyes. He starts by joking about Toby, "I hope he did not scare you off with all his complaining. He tried that with me a couple of years ago. It didn't work, as you can see." They move along the corridor. "Stairs OK?" he asks and she nods. She is grateful for a different kind of activity, even for a few minutes. Walking up a few flights of stairs, Chloe notices that Barry is not only energetic but also athletic.

Before they even get to his office, Barry is well into his questions. How was her day, what did she think of Matt, the chairman, always the first meeting of the day? Boring, no? And what about Peter, Cathy and Hans? Toby they have already finished with. He talks as if they are co-conspirators. It feels very different from the other conversations of the day. She senses a slight edge to some of his banter. Or perhaps she is misreading it in her tiredness. She makes a mental note to be careful, not to fall into the trap of saying anything she shouldn't. After all, Barry already has a job here. Some of his comments about the other faculty members are quite funny, though. Matt is a bit boring, it is true. He initially calls Hans Meiner the old man, but quickly modifies it to the grand old man. Peter gets a tougher treatment. Barry does a good mimic of Peter's leaning-forward earnestness when discussing a new area of interest. You'd think he personally discovered each new field that he gets into, Barry says. There is some truth to it, Chloe thinks, but it is also a bit harsh. She lets it slide, ready to have a more relaxed conversation with someone who clearly wishes to entertain. She has noticed that the American discomfort with irony and sarcasm seems even more pronounced here on the West Coast. But with Barry, she thinks, she need not worry.

Sitting in Barry's second floor office, they finally exhaust the small talk about her day and her impressions and switch to discussing science. He asks a few questions about her talk and future plans. Like Peter, he is interested in the normal regulation of Jmjd10 and speculates about the nature of its membrane association and release. Chloe discusses her ideas and plans for addressing the question. In part, this is a repeat of her earlier discussion with Peter. But she finds that she has expanded her ideas and improved the argumentation. It is odd how the mind works, she thinks, how explaining something, verbalizing, seems to move things around in your brain. New ideas present themselves, even if you have thought about these very same issues numerous times before. She knows that this is nothing new, but experiencing it herself, feeling the ideas well up, organize themselves in her head, excites her. Barry reciprocates by telling about his own work. He works on membrane lipid-protein interaction, which helps explain his special reverence for Hans Meiner as well as his interest in those specific aspects of Chloe's work. She finds the work fascinating and does not have to try hard to keep a good discussion going.

Later, when they have already passed their allotted forty-five minutes, they talk about living in San Francisco. Like Cathy, Barry is very enthusiastic. Chloe relaxes further and starts to notice the details of the office she is in. Except in Toby's office, she has paid no attention to her surroundings today, she realizes. In Barry's bright space, she notices two large Ed Ruscha prints. It gives her an opportunity to tease him a bit.

"For someone who is supposedly so fond of San Francisco, you seem very LA" she says, nodding toward the striking prints. One is a nighttime cityscape, criss-crossing streetlights, the other is a mountain with the usual incongruous text that she reads but then immediately forgets. "Is that where you are from? LA?"

"No, no. I just like Ruscha. They give me perspective—or something. I'm from New York and from England, a transatlantic hybrid." He pauses. "I went back to New York for my postdoc. I like San Francisco and New York—and I seem to belong to a small minority—single, non-gay men. And I take full advantage of this." He adds with a slightly roguish smile. "Do you like to go out? Or does your hubby keep you at home?" This is way too personal for the situation. But she decides to answer it anyway. She already told Matt that there was no 'two-body problem'. In other words, they do not need to worry about a 'significant other' finding a job as well. They just need to decide about her.

"No hubby." She says. "But I am afraid that I am very boring. I work too much."

"Boring, schmoring. We all work too hard. It goes with the territory. Let's do something fun then, before the serious dinner with Matt tonight. We'll be going to dinner at a nearby restaurant they always seem to take job candidates to. Not too exciting." He notices her expression. "No, it's fine, don't worry. It's a good restaurant. I'd just appreciate a bit of variety for my selfish self, that's all. Anyway, it is relatively close by. So if you don't desperately need to go back to the hotel, we could stop by a nice little bar that I know on the way. What do you think? An hour alone at your boring hotel or an hour with me?"

"A stroll sounds perfect, thanks." She answers with a short laugh, amused by his antics. "And I could manage a drink at this point."

"Excellent. We'll just run by Matt's office to pick up your bag and tell him where we kids are off to. Matt and a couple of the other departmental heavy hitters are coming for dinner. They will want to grill you some more. So let's fortify you first." He seems to enjoy including her in this game of 'us versus the old guys'. She doesn't mind. It also seems like a good plan to have a de-stressing transition between all that has happened today and the dinner, where, she now realizes, she will have to be on her toes as well.

The bar that Barry takes her to is low-key, lively, but not too noisy to allow a conversation. It seems to cater largely to students, but with a selection of mature faces as well. She orders a glass of local Merlot.

"The city is provincial compared to New York and London." Barry admits, "But then, everywhere is. It should be a step up from Boston, I would think." Chloe feels no desire to defend Boston's honor. She chose it for the science and feels no strong attachment to the city. Moving on to their interests and backgrounds, they find plenty of common ground, from being only children to enjoying modern art from the safe distance of non-experts. Barry tells a bit about his boarding school years. Imagining being sent away to school at such a young age, Chloe finds her minor grudges against her demanding father and her self-effacing mother slipping away. She tells him about the challenges and pleasures of growing up multi-lingual. Bit by bit, the conversation takes turns that are utterly different from what she had expected. The banter is left behind and she glimpses other facets of Barry. It is with some regret that they realize they have to be getting on to the restaurant. Chloe prepares to put her strong face back on and Barry slips back into light banter. He returns to the subject of Matt and his lack of imagination in choosing the restaurant for these job interview dinners.

As it turns out, the restaurant is ideal for tonight's purpose. It is a small Italian restaurant with friendly, non-intrusive service. There are enough semi-private nooks and crannies that even their table of 6 has unhindered conversation. If Chloe felt a bit old in the bar, she feels like the youngster taken out for dinner here. Barry has retreated somewhat now that the more senior faculty members are there. He probably appreciates that she alone should be the center of attention tonight. Slightly relaxed by the wine, she feels comfortable but she drinks only a small amount more as the dinner progresses. This is work, after all, and she has to think straight and project well.

The early part of the conversation focuses, as expected, on her work, the talk, the recent paper, her future plans. This part comes easily to her, the questions predictable from several rounds of interviews and dinners. The only unpleasant moment is when Gunner Nüberg speaks up from his otherwise quiet corner. He feels that she should have cited some earlier work of his in her recent paper. Chloe is not sure how best to defend herself, not knowing his work well enough to come up with convincing arguments. She is saved by Matt and the two other department professors. They claim not to agree with Gunner and make jokes about his citation record surely not suffering too much.

It is at this point she understands that they are actively trying to recruit her. It is not just her trying to impress them. It is also the other way around. This comes as a sudden revelation. Yes, both Cathy and Barry expressed an interest in recruiting her. But this is different. This is the chairman and others with, she assumes, a significant say in the hiring. She lets this sink in. The topic of conversation shifts to Tom and his lab. They all know him, at least professionally, and are looking for standard updates. She obliges. She mentions, in passing, how happy she is that Tom encourages his postdocs to pursue independent ideas and that he lets them take their projects with them when they leave. She figures there is no harm in making this clear. The topic of Tom is relatively quickly exhausted, and the rest of the dinner conversation is less focused on Chloe. She can relax a bit. They have done their job, all of them, and they are just passing the time now. Matt finally calls for the check to make it an early night. "Another busy day for you tomorrow, I am afraid." He says to Chloe. She is aware of this and vaguely relieved that the dinner is over. It is good to wrap it up before she gets too tired to think properly. She needs a solitary hotel room, a bed and some sleep.

Chloe is lying on the bed, still clothed in her carefully chosen job-interview outfit: a loose but elegant shirt with narrow stripes and black linen pants, discreet clothes but still her style. She is exhausted, but too wound up to sleep. She knows she will spend the next hour reliving the day, interpreting and evaluating every aspect of it, every conversation. She did the same after each of the previous interviews. The very first one, back in December, in New York, was a disaster. Well, perhaps not a disaster, but not satisfactory, not at all. She remembers it all too clearly. Coming to the medical school by the hospital entrance started the whole day wrong. Then the dingy classroom for the seminar, only half-full. She got no traction from the questions after her talk, all about medical relevance, odd technical details and some even wholly irrelevant to her work. The ridiculous situation with the famous Professor Park still infuriates her. He hadn't bothered to come to her talk, but deigned to allow her to entertain him in his office while he almost fell asleep. Arrogant bugger. But what bothered her most about the interview was not that the place failed to live up to her expectations. No, it was her. She did not perform well. She let herself down. Nerves got to her and made her talk a fumbling mess.

They all said they liked it, but she knows it was not great. The one-on-one interviews with junior faculty were OK. But with the senior ones, awkward. The session with Sandra Howard was particularly terrible. She could have kicked herself. Knowing Sandra's exquisite work, she wanted to make a good impression. But the conversation never sparked. Instead Sandra went on and on about the importance of another female hire. Chloe hates that. She wants to be respected, and hired, as a scientist, not as a female scientist. But she could not have that discussion with Sandra. No, she did not do well there. New York was nice, though, and she had fun at the dinner with some junior faculty and the chairman. But she knew she had blown it. She probably wouldn't even want the position, in that strange department. Nevertheless, performing suboptimally bothered her a great deal. No way she was going to do that again. So she worked hard on improving every aspect of her performance and interactions in the following interviews. And it got better each time. Today, everything went as well as she could possibly have hoped. But boy, was it draining. The one weak note was her inability to answer Gunner Nüberg's challenge at dinner. She decides to look up his papers tomorrow before breakfast, find the one he was referring to and be well prepared for their chat later on. Then she can defend herself properly while showing him the respect of knowing his work. That should do the job. Satisfied that she has no more issues to iron out, she forces herself upright to get properly ready for bed.

Next day, Chloe feels well rested, tuned and ready for anything. She is ready to face another day of concentrated one-on-one interviews. She is ready to listen and to be sharp. Bring it on.

The final 'wrap up' discussion with Matt takes an unexpected turn. They talked loosely about the department, university facilities, graduate programs and so on yesterday morning. Today, he asks how her day has been and reiterates his admiration for her talk and the recent paper. He has also read the article in the New York Times. Chloe keeps forgetting about Frank's article. It was published just after she left Boston for this long trip, so it drifted quickly into the background. There was too much else to think about. But it has been noticed, it seems. Matt finally goes quiet and clears his throat. She waits, intuiting that he has something important to say.

"This morning, I have spoken to all the other members of the search committee and a few others. I want to tell you that we are all very impressed with you and your work." He leans forward a bit. "We are very keen to recruit you here, very keen. With your focus regulation of cell death and cancer relevance, you would fit in perfectly. We really hope we can get you to join our faculty." He pauses briefly. Chloe dares not interrupt. "We do, however, still have a few more people to interview, scheduled for the next month's time. So unfortunately, I cannot offer you the position here and now. But I wish I could. I understand that you have had successful interviews at several excellent places. So you will probably have choices. But I want you to know where we stand, so you leave us with the right impression. If you are interested, that is."

This almost-offer is unusual at this stage; they both know this.

"Absolutely, I am very interested" Chloe answers immediately. She does not have to think about it. "I really enjoyed my time here. I had a lot of great discussions. And of course, the reputation of this place is formidable. So yes, definitely, you can count on me being interested." Matt seems pleased and proceeds to tell her more about various practical plans for the department and the university. She doesn't retain much of this. Her voice did not betray her just now, but she is in fact stunned. Happily stunned. She has done it, landed herself an independent position at one of the best universities in the US. And she knows it is a place where she really wants to be. Well, almost done it, she corrects herself. She is not quite there yet. But it is impossible to contain the joy of what has just happened. She nods and smiles at what Matt says. Perhaps she should be playing it cool. Later, she can do that later, when she has the job-offers. Right now she has to struggle just to keep from grinning too wildly, like some jolly lunatic.

Chapter 9

It is time. Past time. The discussion with Tom was scheduled for 9 A.M. so he will be expecting her by now. Karen has been at her desk since early this morning, but has not been able to settle down to do much. And now she is late. They are supposed to talk about how to bring her work to publication, as a continuation of yesterdays group meeting. That part is pretty straightforward. But she cannot stop thinking about what Andy told her regarding Chloe's mice and the numbers in Figure 7F. Should she tell Tom or should she let it go, forget she ever noticed? She has to decide, somehow. Is she obliged to say something? She doesn't know. She also has no idea how he will react. What will he think of her for bringing up such a suspicion? But it is time to get going, it is. After a few extra minutes of hesitation, she pulls herself together.

She picks up the laptop and walks slowly through the lab, into the corridor and the few steps to Tom's office. The door is open and there is no sound from within. No last-minute phone call or urgent writing to free her from her predicament. She takes a deep breath and steps inside as she taps gently on the open door. Tom looks up, smiles encouragingly and gestures to the usual chair. When he speaks, he switches to a more serious expression.

"So, Karen. That was terribly bad luck with the Science paper. And so unexpected. I've never heard of this group before. You did have all the right ideas and yours would have been a much better paper. That was clear from your presentation yesterday."

She is grateful that Tom refrains from mentioning his advice to write up her work earlier; she needs no 'I told you so'.

"It was a total shock." She says. "It has really thrown me off. I have to admit that in the last few days I have been more than ready to give it all up, to call it quits."

"That is completely understandable. You are not too thrilled about science, or your career prospects, right now. But promise me not to make any big decisions immediately, OK? You will get past this. You are very talented, Karen. It is important that you know that." Tom leans forward and looks her directly in the eyes. "Not everyone who comes to my lab has what it takes. But you do. You have the smarts and the tenacity to make it in science. I can see that in how you run your project. And you have a mind of your own, I don't have to feed you ideas." Tom smiles encouragingly again.

Karen's surprise at his words shows on her face. She didn't think that he had really noticed her. And yet he seems so certain about her abilities. As much as this pleases her, she is momentarily unable to come up with an adequate reply.

"So don't give up, OK? You have a good independent career ahead of you, if you want one." A short pause and he continues, "Now, let's sort out how you can pull together a story and get it published this year. It's still January, so there is a good chance. Then you can start thinking about where you want to take your project afterwards. And don't worry too much about your fellowship running out. I can extend you so you leave here with another paper under your belt. You can do it. You just need to get over this set-back." Another pause. "So, what do we need to wrap up this story? Not too much, according to your presentation yesterday. You probably have the first four or five figures already. You will need to add data on the difference between transformed and normal cells. You mentioned yesterday that you had some results already. Did you find them?"

"Sure, I .." She opens her laptop. But before she is able to say more, he continues.

"And, as you suggested yesterday, it would be excellent to try your 3D culture experiments, visualizing cell connections in normal and transformed cells, with the cell death sensor transduced in as well. And finally, it is certainly worth trying the mouse experiment that Vikram suggested. So let's go through these parts, one by one. OK?"

Turning to itemized practical issues untangles Karen. That, and Tom's encouragement. She finds herself addressing the problems with more focus and enthusiasm than she has been able to muster for quite some time. They look at the figures she has put together already. Tom has a few suggestions for changes, but they do not dwell on it for long. They discuss the data she has so far on the transformed cell lines. It is a useful discussion for her, giving her a better idea of how to write the corresponding result section. It also allows her—or them—decide which cell lines to try in the more difficult 3D experiments. These assays are labor-intensive and require lots of microscope time. So making the right choices upfront is important when time is tight. Finally, they get to the mouse experiment.

"So did you talk to Andy? Does he have mice available for you?"

"Yes, I did, and luckily, he does. He has plenty of both genotypes, so we can set up enough matings to get everything going quickly. Even with the small litter size, I should get enough double transgenic pups. Maybe even enough to do several time-points."

"Ah, very good. I remember working with these mice. The tumors come really fast. So it should be possible to get these experiments done within 3 months, and the others, too. So we have a timeline to aim for."

"I think so. I'll get the virus preps ready and I can start checking the imaging conditions. Then I will be all set when the transgenic mice are ready. And the equipment downstairs looks OK, I checked. In the best-case scenario, I will be able to image the live mice. If not, I'll have to make explants of the tumor tissue. But that might also be OK. The only issue is whether the signal will be strong enough for me to see 'nanotubes', or whatever, in the tissue. I will just have to try."

"Well, it sounds like you have it all under control. And I'm sure you will give it all you've got." He pauses. "So, nomenclature. We should get used to calling them nanotubes, I suppose. That is, unless you are convinced the structures you are looking at are not the same as those in the Science paper."

Looking at it objectively, Karen can see that her 'thin connections' are almost definitely the same structures as the 'nanotubes' described by the Polish group. She explains why to Tom, as well as how she can experimentally confirm it. He agrees. This being the case, it will be best to use the same term to avoid confusion.

Tom throws a glance at his computer screen and Karen starts to get up. But she hesitates and sits down again.

"There is something else. Something that I should probably tell you."

Tom looks up. Karen is still unsure. Maybe she should leave it well enough alone. No, it has to be Tom's decision.

"It is something Andy said, when he was looking up the mice for me."

"Yes?"

"He said the reason he had so many mice now is that he thought Chloe would need more. But she hasn't asked for them."

"Of course not. Her experiments were finished almost half a year ago. She got what she needed for the revisions."

"Well, I know, but" she hesitates again "Andy and I were talking about how many mated females I would need for my experiment. This led us to why Andy had all these mice ready, even though I had not preordered them. He said he was expecting Chloe to want them because he only gave her two pregnant females last summer. He said she was very annoyed that he did not have more mice for her at the time."

"Well, presumably she didn't need more. Where is this going?" Tom's tone of voice has hardened. She realizes that she should not have mentioned this. But having started, she will have to finish. She will have to say what she has to say and then get out of the office as quickly as possible. It should have been obvious to her that Tom would not want to hear this. He may not take kindly to the bearer of the information, either, whatever the explanation may turn out to be.

"Well, Andy also said he thought one of the two females had died. And that leaves just one. I looked back at Chloe's paper." Tom has been focusing on a piece of paper on his desk. Now he looks up at her with an unreadable expression. She blunders ahead. "And I saw that she had used six double transgenic mice for the last experiment. Three with, three without drug." She pauses, wondering if she needs to say more. But Tom is silent. "She could not get six double transgenic pups from one pregnant carrier female. Even with two females, it would be very, very unlikely. Andy says the litter size is small."

Tom remains silent. After a while, he answers coolly, and with seemingly deliberate calm. "If the paper says six double transgenic mice were used, then six were used. Obviously."

Karen does not know what to say to this. "Sorry. I just thought you should know." She gets up and gathers her things quickly. On her way out the door, she adds "I'm not...", but does not finish the sentence. Instead she closes the door behind her, firmly but quietly.

Tom stares at the papers on his desk, not really seeing them. He needs to think carefully. If Chloe said that she used six mice, then she used six mice. Why would he doubt that? But if what Karen said is true, then something is way off here. "Damn her" he says, out loud. He is not sure if he means Karen or Chloe. Maybe both. He gets up from his chair and starts pacing in the small office.

He knows these mice produce small litters. If Chloe only had one pregnant mouse, that would never work. Even with two, it looks fishy. But he cannot believe that she would fudge data. It cannot be. There must be some other explanation. Most likely, this is some sort of a misunderstanding with Andy. Karen shouldn't be meddling. Maybe she is jealous of Chloe's success and trying to stir up trouble. He has seen healthy competitiveness turn into unhealthy envy plenty of times. Karen is struggling seriously right now and her outlook is uncertain. She could be prone to a sour turn, whereas Chloe has a lot going for her. It shouldn't matter, really, but he thinks Chloe's charm also brings out envy more easily, especially in other female postdocs. It is just human nature.

But why now? Why would Karen tell him now and risk poisoning the constructive interaction they just had? This does not make sense. Envy and spite work best in the dark. Karen could just quietly spread the rumor. Maybe something is wrong. The experiment with the double-transgenic mice was the very last experiment they needed to get Chloe's paper accepted. They were pressed for time, a very stressful situation. It is possible Chloe crossed the line. No, there must be a reasonable explanation. He must talk to Chloe and sort this out. But that means waiting for her to get back. But he cannot wait. It has become his problem and he needs to do something now. People will be quick to blame him if there is any misconduct in his lab. They will be double-quick and double-harsh if he appears to be hiding it, or refusing to consider it. He knows a few competitors who would love to see him get into this kind of ugly trouble. If they found out what Karen has just said, and he did nothing, they would be after him with a vengeance. That he himself has done nothing wrong is not the point. It is his responsibility. He should talk to Andy right away and see what he has to say. Maybe he can put this thing to rest. Talk to Andy and then decide what to do about Chloe and what to do about Karen. He should probably talk to someone else as well. Someone who would know how to handle this situation, if it is a situation. Maybe Stuart. He is a sensible person and Tom knows he has been on the university ethics committee. Plus he is a good friend. Lunch—he will ask Stuart to go for lunch with him at the faculty club.

Having come up with a plan for how to proceed makes Tom feel marginally better. He will not let himself dwell too much on the possibility that Karen's suspicions are correct. They cannot be. The paper has just been published. Chloe is out there giving talks. It must be a misunderstanding of some sort. Chloe is too smart to do something like this and jeopardize her whole career. He will not cross that bridge unless he has to. So first Andy, then Stuart.

Half an hour later Andy is at Tom's door. He looks ill at ease, unused to coming to Tom's office. Tom invites him in and closes the door for privacy. This makes Andy more nervous.

"Andy, you are not in any trouble." Tom starts, "I just need information about some mice Chloe used last summer." He emphasizes how important it is that he gets the correct information and then brings up the specific issue. Andy is quiet for a moment, then repeats to Tom exactly what the told Karen. The cross was MMTV-Myc transgenic females crossed to MMTV-Ras males. Both transgenes are kept in heterozygous state. He gave two pregnant females to Chloe. And he is pretty sure that one of the females died.

"But I wouldn't have that on record." Andy explains. "Once I passed the pregnant mice on to Chloe, record keeping would be her job. But the two plugs, I mean, the two successful matings of MMTV-Myc females, those would be on record. The rest is just how I remember it. I still keep an eye on the mice, sometimes, even when someone else has taken charge."

Tom looks unconvinced. Andy explains further. "Last summer, Chloe got very angry with me when I told her how many mice I had available for her. There were just the two females that she could use immediately. I really don't like being yelled at. That is probably why I remember so clearly that one of them died. I was sure she would go ballistic again and blame me, even though the mice were now in her care. She didn't, though. She didn't get back to me at all. I expanded both strains as soon as I was able. I figured she'd want more later on. I don't know anything more than that."

"And you say you have records of this?"

"Well, we can look it up. You should have direct access to the database on the mouse-house server, via the internal network. But it would just be my records, until I hand the mice off to the investigator. So we should see the two matings and the date of the plugs."

Tom turns his screen so they can both see and Andy guides him through the logon to the database. They find the relevant files easily. As Andy had explained, the files show that two pregnant females were given to Chloe last July. Listed are dates of the plugs, full genotypes of females and males used and the date when they were given to Chloe. There is no further information about them. There are brief entries below showing the two mouse populations being expanded. The final entry is yesterday, the request put in by Karen.

"By the way, you need to OK that one." Andy points out.

"Sure, of course. I got an email with a link to do that. But what about these pregnant females from July? Is there no record of how many pups they had and their genotypes?"

"Well, the arrangement we have is that the postdoc, in this case that would be Chloe, does the genotyping. I discussed this with you previously, several times. We don't have manpower and time to do all the genotyping for everyone's experiments downstairs. We have a pretty big colony to run, as it is." Andy sounds defensive.

"Sure. That's completely OK. I just thought maybe the information would be in the database somewhere."

"No, they are out of the system at that point. But as I told Karen yesterday, don't count on a big litter. You'd get max eight live pups from these mice, on a good day. Anyway, you'll have to ask Chloe. She should have the data."

"Ok, thanks. I'll talk to Chloe about it when she gets back."

"Anything you want me to do?" Andy asks.

"No, not for now. . ... Well, actually, yes" Tom corrects himself "So both the MMTV-Myc and MMTV-Ras strains are expanded now?"

"Yes, I haven't culled any of them yet."

"Good, keep it that way. And keep setting up more intercrosses. Whatever Karen requested from you yesterday plus more as they come along. Just in case we need them. And keep me posted about them." Tom hesitates, then adds, "Oh, and please keep this discussion confidential for the time being."

"Sure, no problem." Andy is out of the office quickly.

Not good, Tom thinks. This did not sort anything out, unfortunately. What he now knows is that this is not something Karen made up. She was just paying attention to what Andy said and she put it together with what she had seen in the paper. She would have had the paper handy, or in mind, because they talked about it at group meeting yesterday. The choice of speaking up could still have an element of spite in it. But probably it was otherwise quite innocent from her side. This does fit with her telling him directly. He should not fault her for that, he knows. She did the right thing. So now he will have to ask Chloe directly. This will not be a pleasant conversation. Even voicing this kind of suspicion shows mistrust. She will surely be offended, even if she can resolve it. He has never had this kind of situation in the lab before, thank God. And then Chloe, of all people. She is so super-talented. He will have Deidre check out her travel plans so he knows for sure which day to expect her. Now he should wait and see what Stuart has to say at lunch. Tom settles in to try to get some work done before then. What he needs is a defined job to distract him from this potentially explosive mess. He started reviewing a manuscript yesterday. He should be able to complete that.

The faculty club is small but pleasant. The view is not interesting, dominated as it is by the massive gray university buildings. But the inside space is attractive, in keeping with the rest of the building. Tom acknowledges that it is an indulgence to have a faculty dining room for the size of the institute. But it is convenient at times like this. He stands near the entrance, waiting for Stuart. Selected pieces of artwork from the institute's collection are on display nearby. To occupy his mind, Tom studies the new pieces, new since his last visit here. The paintings and lithographs are all originals, he thinks, probably donated by wealthy trustees. He imagines well-meaning trustees thinking that extraordinary science might be cultivated by subtly inspiring the scientific minds with other types of creativity. A nice thought, but not one that he actually believes in. Maybe he is too pragmatic. They are interesting people, though, the trustees. The ones he has met have been intelligent and open-minded. And the artwork does lift the place up, whatever its intent.

Stuart shows up on time and is his usual gregarious self. He is about Tom's age, mid-fifties, short and compact, with curly dark hair, heavily infused with gray. He may be called distinguished, but not handsome. There is too much of the troll in him for that. He can talk a mile a minute, but is also a good listener, as Tom well knows. That plus an uncanny knack for knowing all the latest campus and science gossip, means Stuart is always good company. Tom and Stuart have known each other for decades, since their postdoc days.

They are shown to a table near the windows and quickly order today's specials. This is private enough, Tom thinks. The small talk is cut short soon after they are seated, as Stuart gets to the point.

"So what is this about, Tom? I'm always happy to have lunch with you, of course. But Deidre told Sue that you wanted to meet today. So I assume there is something urgent?"

"Well, yes. There is. I need some advice. I have a tricky situation in the lab. Or I may have a tricky situation in the lab. I'm not sure yet. But we could be talking about misconduct. Just maybe, nothing is clear. It concerns one of my postdocs. I just heard about this this morning so it is all very fresh. I wanted your take on it, informally, you understand. It may all turn out to be nothing. But I'm afraid that something is off. So I thought I'd run it by you and see what you think before I do anything."

"No problem, Tom. I'm listening. Possible misconduct. That can be a real minefield, these days. It will stay between us, if that's what you want."

Tom starts to tell the story of the day's events, giving some extra background about the Nature paper. Stuart might have noticed the paper, but he is unlikely to have read it. It is not his area, after all. Tom explains that today's issue only concerns one tiny bit of a large body of work. He tries to give Stuart a fair impression of Chloe, as well as of Karen and Andy. He insists that none of this makes sense to him, as he trusts them all. He finally concludes that maybe he is overreacting.

"It could all just be one big misunderstanding." Tom repeats. "I haven't talked to Chloe yet. She is on an extended West coast job-interview tour. And she is doing extremely well, I hear. She is very, very talented. I know I have to talk to her but I am dreading it. Just asking about it means I consider her capable of doing this kind of thing. Anyway, when we were revising the paper, she showed me the numbers for the experiment with the six mice. I didn't actually go down to see the mice. Why would I? Asking Chloe about the numbers now would be suggesting that she was lying then. And I don't want to imply that unless I am sure. But I am not sure. I can't be sure without seeing the relevant raw data, which requires talking to her about it. What a mess. Maybe she was just very lucky and maybe Andy is misremembering about one female dying."

"But even with two females you said that getting six correct pups was unlikely." Stuart interjects.

"Yes, but it is not completely impossible. It is just very unlikely. And maybe Chloe had some extra mice somewhere. I can't know for sure whether there is anything wrong at this point. Maybe I shouldn't even be discussing it before I've had a chance to talk to her. But I have to make sure I do the right thing here."

Stuart is quiet for a while before he responds. "Look, Tom. This is a bad situation no matter how you look at it. I imagine you wish it could just go away. But that is not an option. This is a more-or-less classical whistleblower situation. Now that it has landed in your lap, you have to follow up on it. As a colleague, and as a friend, I have to tell you that it can harm you and your reputation, and I mean seriously, if you do not do the right thing, and immediately."

Although not unexpected, Tom winces at hearing Stuart's words.

"I guess I knew you'd say that. That's why I wanted to talk to you. And what is the right thing, then?"

"Well, first of all, you should contact the director's office and explain the situation. They liaise directly with the office of integrity at the university. There is a lawyer there who they normally use for situations like this. She is competent and sensible, and she knows a good deal of science. I have dealt with her on several occasions. Nayar, I think her name is, Sushma Nayar."

"But isn't that a little over-the-top at this point? I mean, it could still be a misunderstanding of some sort."

"Right, you should tell Sushma that, of course, that nothing is clear. But she or one of her people in human resources should be present when you talk to Chloe. To make sure everything is done by the book and to do what needs doing for follow-up. In a way, it is good that Chloe is away, so you have time to set this up. When will she be back?"

"Soon. In a couple of days. But is this fair to her? If I do it this way, without hearing her side first, it will seem like I have already judged her. At the very least it will tell her that I suspect her of wrongdoing. And I'm not sure about that. She is ambitious. But I've worked with her for over 4 years and she has always seemed completely straight. She knows the rules. I can't imagine that she would risk everything by doing something like this. There is too much to lose. I should talk to her quietly first, see if is possible to sort it out without any fuss. I don't like springing these 'office of integrity' people on her without warning."

"Look, Tom. I've seen situations like this before. Not a lot, but it happens. Someone accuses someone else of misconduct—or just points to the possibility of it—and it has to be sorted out correctly. What you can do is to first explain to Chloe why you have to do what you are doing. Explain that you have no choice but to follow procedure. She should also be told that the procedure is confidential, so no one on the outside will ever need to hear about this, if it is unfounded. But it has to be cleared up by a neutral party. So you should have Sushma or someone from her office ready to step in immediately."

"But what if I just try to sort it out with her first? To me, that seems the most natural way. It is my lab, after all."

"Well, exactly, it's your lab. And a major paper with your name as corresponding author rests on this work. You don't have to be too cynical to see that you have a vested interest in the outcome. And that's why it's essential that you don't handle this alone."

"Vested interest?"

"Come on Tom, don't be dense. You will look bad if the paper has to be retracted, so you can't be neutral about the outcome." Stuart says in a tired voice. He should not have to explain this, but apparently it needs to be spelled out. "Tom, you may not like it, but you protect yourself from possible accusations of cover-up by ensuring that the process is under independent scrutiny from the outset."

"But I'd never do anything like that." Tom insists, visibly upset. He knows that what Stuart just said about the paper is correct. But it stings, nevertheless. "I would not sweep it under the rug. I know that is much worse. That is why I am here, talking to you."

"I know that, Tom." Stuart says in a gentler tone. "Of course I do, but you asked me here so that I could remind you of the correct way forward. And in this case, full openness and evaluation by neutral party is it. That way, no one will ever be able to claim that you were anything but completely straight, whatever happens. The two of us talking now is off the record. It doesn't help you in any official sense. By involving the director's office, you put yourself in the open and hopefully in the clear. If things get worse from here, you'll be glad to have done it this way."

"I suppose I see your point. But I hadn't thought it through, not completely. Chloe is one of my best postdocs ever. Ambitious and all, but her work is really impressive. And it's a really good paper. I'm not just saying this because it's from my lab. It's solid. We've looked at this stuff from so many different angles. All the experiments hold together. It all makes sense. I'm sure the paper is correct."

"Well, maybe nothing is wrong and nothing more will happen. Chloe will sort it out with Sushma and the director's office. And yes, she will be pissed off at you for calling them in. That is unavoidable. But if she did nothing wrong, she'll get over it. Eventually. And later, when she has her own lab, she will understand why you had to do this."

"You are right, of course. The voice of reason. Thank you for that, my friend"

"We all need a prod, sometimes."

Tom's fish has remained untouched. He makes an effort to eat some of it, but it has gone cold. He pushes it aside.

"Coffee, Stuart?"

"Yes, of course."

Tom is both more and less troubled than when they started out. On one hand, the possible implications of the whole thing are now more tangible, unpleasantly so. It could end up being very bad. On the other hand, it feels good to have talked to Stuart about it. Unburdened, he is calmer. And if it all unravels? He will just have to deal with it, somehow.

Stuart lets Tom ruminate. He has said what needed to be said. He knows it, but he does not like having to push a friend in this manner. When the coffees arrive, Tom speaks again, his tone now resigned and philosophical.

"It's remarkable, how fragile it all is."

"Fragile?"

"Well, at least in our area of science, everything depends so much on trust. I have to trust the people in my lab. If they show me a number, I need to trust it is correct. In papers, you have to trust that people present the correct facts. Otherwise nothing would make any sense any more. Maybe in other types of experimental science, like particle physics or whatever, it is different. It's more of a joint effort for them. The raw data are seen by many eyes. So everyone would know when something wasn't kosher."

"Well, that type of science has its problems too, you know. Maybe other areas just look simpler because we see them from the outside. In molecular biology, we rely on whether something can be repeated as the litmus test. We present the data so that other people can re-test them. Lots of labs can check just about any published result. You don't need access to a supercollider for that."

"But that comes after, and it doesn't prove much one way or the other. It doesn't prove intent or misconduct. Anyway, no one has the time or the inclination to test someone else's results."

"Not unless they really have to. If the new findings contradict their findings—or if it is important and controversial enough—they do. In this way, mistakes do get weeded out eventually."

"Yes, I know, but trust is still fundamental. I can't help thinking of the journals that look so carefully to see if any images have been manipulated. But they still have to trust that the experiment is what it claims to be and trust the numbers. It is just strange, in a way. Our world is extremely competitive and we all know that. At the same time, it all depends so much on trust."

"Well, it is built on trust. I happen to think that most scientists are trustworthy. And then there is knowing that you risk losing everything if you ever misstep. A little extra incentive for the wavering ones."

"Right. Anyway, enough of that for now. I will get on with it as you suggested." Tom picks up his coffee. "Now, tell me about Josh. Has he decided what to do with his life? Will he follow the parental paths of fame and glory?"

Stuart smiles and shakes his head.

Chapter 10

Damn that cold wind. Chloe wraps the scarf tighter. After several weeks in California, she is not prepared for the Boston winter. She did not notice the cold as much yesterday afternoon when she came in from the airport. It was sunny and nice. This morning is bitter. Martin did suggest that she go in with him, as he was driving in anyway. But she wanted to go home first. Supposedly, this was to unpack and get changed. She also needed a bit of time alone. With Martin, last night, it was complicated, awkward. They arranged to see each other again tonight, given all her time away. She is not sure it is a good idea, but it was hard to say no.

What she needs, she knows, is to get her head around being back in Boston again. It is not easy. It was all keen focus and eager anticipation before she left for California. All the work she put in on her talk following the botched first interview was directed at doing well on this trip. Even the experiments she started months ago were planned to have preliminary data ready for discussion at the interviews. Now everything feels strangely anticlimactic. She has done as well as she could possibly have imagined at the recent interviews. At least one job offer will come from this trip and, she thinks, probably more. All she has to do now is wait for them to contact her. But she does not feel like waiting. She wants to move ahead, now. She should be in a new space, starting up all the new things, seeing new people. Coming back to the institute, to what now seems like "the old lab", feels like regressing.

As she approaches the familiar red building, she stops up and looks, and sheds the minor irritations. It is still wonderful, still a great place. Yes, a bit distant, now. Soon it will be part of her history. Sad? No. She is not yet ready for nostalgia, just ready for the next stage. But right now she is also looking forward to telling people about the interviews, especially telling Tom. He will be happy for her, she is sure of that. One of 'his people' doing well reflects well on him. There will be no need to hold back or play down the excitement, as she found herself doing last night. She picks up the pace. Tom should be in by now.

Tom's half-open office door shows her that he is in and that interruptions are allowed. She knocks and immediately steps halfway inside, a barely contained happy grin on her face. Tom looks up and sees her. "Chloe, come in. Deidre told me you would be back today." His eyes dart around his desk as if looking for something. He moves a few papers around, clears his throat and looks up again "So how was your trip?"

Chloe seats herself in the chair opposite Tom and starts telling him about her interviews, giddy and excited. She talks about the people she met and which ones she had particularly interesting discussions with. Along the way, regards from various colleagues are passed on. She is bursting with pleasure and pride as she talks about the amazingly positive feedback she got. She recounts, almost word by word, what Matt Rosen said, the virtual promise of a job offer. It takes a while before she realizes that Tom is being too quiet.

"What's going on?" She pauses. "Is something wrong? You haven't even congratulated me."

Tom gets up from his chair and goes to close the office door completely.

"We need to talk." he starts, as he sits down again, heavily.

"Sure, OK. What do we need to talk about?" Chloe sounds puzzled.

"While you were away, something was brought to my attention. It is something potentially very damaging and something that I had to take seriously. And it involves you. Well, it is about you."

Chloe straightens in her chair and waits for him to continue, her surprised expression gradually hardening as he fails to do so.

Tom takes a deep breath, finds a pdf icon on his screen and opens the file. She can see that it is her recent paper.

"Someone in the lab needed to make MMTV-Myc, MMTV-Ras double transgenic mice, the mice you used for the final experiment here." He scrolls through the file, finds the figure and rotates the screen so they can see the bottom of Figure 7. "The mouse experiments that the reviewers asked for. So, this person talked to Andy, naturally. Somehow they got around to talking about your experiment and the number of mice used."

Tom stops and looks at Chloe. Her expression remains unchanged.

"And?"

"Well, basically they wondered how you could have had enough double transgenic mice to do the experiment as presented in the paper. So I talked to Andy and we looked at his records. From what we could see there and from what Andy remembers, I am also finding it difficult to understand how you did the experiment." Tom points to the screen, "You have three with drug and three without, so six double transgenics in total."

"Yes, that is correct." Chloe answers slowly, trying to control the first twinge of anger. "Six mice was sufficient. The effect of the drug was very robust."

"Chloe, the problem is that I can't figure out how you got the six double transgenic pups. Andy's records show only two heterozygous carrier females were mated. And he says that one of them died. Double transgenic offspring tend to be underrepresented in live pups. We usually don't get the expected Mendelian ratio of one in four. So with the small litters these mice have, it would be pretty much impossible to get six of them from one female, and very unlikely even from two."

Chloe stares hard at the screen. "Andy must be misremembering." She says. "I had six double transgenics. I don't remember how many females I used. Where is this coming from, anyway? Who brought this up?"

"It doesn't matter who voiced the concern. The facts are all that matter."

"Well, I know what the facts are. I did the work and I got the results, whatever anyone else says. The results are exactly what I put in the paper. And of course it matters who this comes from, if someone is questioning my integrity. It is offensive." Her voice is strained now.

"Who first mentioned it is not something we can discuss. Anyway, Andy backs it up, about the numbers. He keeps track of the mice."

"Well, Andy is not exactly the sharpest guy around here, or the most reliable. And he is probably pissed at me. I got a little upset with him last summer. He hadn't kept the colony in good shape and wasn't doing much to help me. I should have played nice, perhaps, but it was a stressful time. Maybe he is still sore about that. Anyway, I had to do everything myself. That's why he doesn't have the correct records for that experiment. I had the number of mice I said I had." She pauses. "And in my recollection, you were perfectly content with the data when we wrote the paper." She tries for eye contact with Tom, but he avoids it. "This whole thing is outrageous. I did what I said I did." She gets up and in her agitation almost knocks the chair over, turning toward the door. "You think I made something up." She turns back to face Tom. "That's what you are saying, isn't it? How can you possibly think that?" Tom gets up and moves past Chloe to stand in front of the closed door. He is trying hard to keep calm.

"Chloe, I'm sure there is a good explanation for this. But you have to understand that in a situation like this, we have to follow institute and university procedures." He maintains a steady, even voice. "You have been accused of not having the data you presented in a publication. You deny the accusation. Okay. Fine. Clearly, someone is mistaken here. What I believe doesn't really matter. We have to deal with the situation properly. I have to bring in someone from the university's office of scientific integrity to ensure this is done correctly."

They are both standing, neither of them moving. She realizes that he is preventing her from leaving the room.

"Now?" She asks, incredulous. "You want to do this now, as soon as I step in the door? Right this minute?"

"The sooner the better." Tom says, attempting a strained smile. "Let's try to get this sorted out."

He signals for her to sit and moves to the door. What does he think she is going to do, she wonders, make a mad dash for it? She sinks back into her chair, shaking her head in now unobserved disbelief. "This cannot be happening", she whispers to herself. She hears Tom opening the door and conversing in a low voice with someone outside. Deidre, she assumes. He comes back and closes the door again.

"Let's just wait here. Sushma Nayar will be here in a few minutes. She is from the office of integrity. She knows the law and she will help us work through this."

"A lawyer? Here? Now?" Chloe feels her color rise. She stands up again. "What is this, an ambush? You had this all set up before you even talked to me about it? Unbelievable. You've already decided that I did something wrong, haven't you? Without even talking to me. This is totally crazy."

"Chloe, as I said, what I think doesn't matter. It doesn't change how we have to approach this." He gestures her to sit again. She does, reluctantly. "I asked Dr. Nayar to come over because we need to handle this properly and follow correct procedure. An accusation has been made, and we have to deal with it. Dr. Nayar will take us through what we are supposed to do. So we should wait for her and talk to her." Seeing that Chloe is not about to get up immediately, Tom sits down as well. He can see that she is seething, not ready to acquiesce to 'procedure'. He figures that he would feel the same, in her shoes. So he says no more. Chloe folds her arms across her chest and turns to face the wall of books. Tom looks at his computer screen, but the tension in the room keeps him from doing anything useful. A few minutes later, a knock releases them. He goes to open the door.

Sushma Nayar is a striking woman, her features strong, regal. She looks to be in her forties, with subtle gray streaks in an impressive mane of shoulder-length black hair. Her business suit is all lawyer, but offset by a bright pink and yellow scarf. As she greets them and introduces herself, her deep-set black eyes rest on each of them in turn with interest, or even sympathy. Introductions completed, she addresses Chloe in an unexpectedly gentle voice

"Chloe, I know this must be a shock for you. It is a very unpleasant situation. But perhaps I can ease it a bit by explaining what is happening and what we have to do. University policy requires that faculty members inform the dean or the director if anyone brings problems with scientific conduct to their attention, even just the possibility of such problems. I was called in by the director of your institute to act as an impartial analyst." She nods thanks at Tom for clearing a chair and sits down, facing Chloe. "The director uses our office so that we can make sure everything is handled properly and everyone involved is protected appropriately." She makes brief eye contact with Chloe. "These are sensitive issues and it is in your best interest that any allegation is investigated in a fair and transparent way. And you have to understand that a full official investigation may be requested from the University's side. All of this will, of course, be confidential. So that's the formal background. When Tom told me the agenda for your trip, we realized that this morning would be the best time for our initial discussion." She pauses. "I have heard from Tom about the specifics of the situation. Yesterday, I talked to the person who made Tom aware of the potential problem, as well as the mouse technician," she glances at her notes, "Andy, who very helpfully showed me his records. I have also looked at the Nature article in question. I have scientific training, a PhD, in addition to my law degree, so I have some understanding of what this concerns." Another pause. "What I would like, now, Chloe, is to hear from you."

Chloe glances briefly at Sushma and then lowers her gaze again.

"You are probably very upset right now." Sushma continues. "This is completely understandable. No one likes to be suspected of wrongdoing. But we, the administration, and your lab head here," she nods at Tom, "we have to listen to any report that might be construed as evidence of misconduct. And we have to look into it. That is why I have to be here." While she speaks, Sushma bends down to retrieve a notepad from her bag and places it on the table. "We may be able to sort this out right now. If we do, I will report back to the director's office and end it today. This is what I hope will happen. I hope that everything can be explained in a satisfactory manner. Do you feel able to talk it through, right now?"

Chloe nods and looks up. Sushma holds her gaze briefly and then looks at her notes. "So Chloe, we are looking at the experiment reported in Figure 7F of your paper. Three MMTV-Myc, MMTV-Ras double transgenic mice were treated with a drug, and three without, and tumor-burden examined 5 weeks later. What can you tell me about this experiment? I understand that it was done quite recently."

Throughout most of Sushma's monologue, Chloe has been looking at her own hands, clasped in her lap. Her body is rigid with anger. When she answers, she addresses Sushma only, not Tom. She starts by explaining the purpose of the experiment and the set-up. This is probably unnecessary, but she is not interrupted. Finally she gets to the details of the mice and the numbers. "With respect to sample size, I used three double-transgenic mice for each treatment, as I reported in the paper. Of that, I am completely sure. I started with more than one or two females. That would obviously not have given enough offspring. The explanation for all this is quite simple. Andy set up some matings, yes, but I set up more myself. Some were done the other way around, using MMTV-Ras females. They are less healthy, which is why Andy did not want to use them. But with no time to lose, I set up everything I could. I am also certain I had more MMTV-Myc females. Why Andy does not have this registered, I have no idea. Anyway, I was the one doing the experiments, not Andy. When the pups were born, I genotyped them, not Andy. After weaning, I maintained the double transgenic mice with or without drug. They were sacrificed 5 weeks later and I counted the macroscopic tumors. This is the tumor burden that is reported in the figure. The difference with and without drug was so obvious that I did not have to look closer."

Tom is looking at the paper on his screen and adds. "Yes, tumor nodules per animal is what is plotted in the graph."

Sushma waits to make sure that Chloe is finished. "OK, Chloe. Good. So you can explain how you got the indicated number of animals of the correct genotype for your experiment. Just to be sure, you say that there were no irregularities of any sort with this experiment?"

"None."

Tom lightens up. So there is a good explanation, as he had hoped. Andy did not know about the other mice.

"But you are not sure how many females were used and how many pups were genotyped to get the six double transgenics?" Sushma continues.

"No, not off the top of my head. I would have to look at my notes."

"Right, notebooks, yes." Sushma makes a few brisk notes before she continues. "So all we need now in order to settle this is for me to have a look at the raw data from this experiment."

"The raw data?"

"Yes, of course. The data you used to make the graph."

"They are just numbers. The numbers are in my notebook."

"But there must be something else. Some sort of evidence that can prove you had these results in hand when the paper was submitted."

"No, just the numbers. Don't you need evidence from my accusers? You just believe them? Why do I need to prove I'm innocent? That doesn't seem right."

"Chloe, this is not criminal law. This is about rules of conduct in science. I'm sure you know all this. It's covered in the course on ethics and good scientific practice that you took when you started here as a postdoc." She looks briefly at Tom, who nods affirmatively. "You have to keep proper records of your experiments, in hard bound notebooks, of course. You also, very importantly, have to keep the supporting data files. That way if anyone ever questions your results, you can back them up with hard evidence. So you see, this is not about whether we want to believe you or someone else. Obviously, Tom would be delighted to believe you. But this is ultimately about what you are obliged to be able to document, when questioned, and only that." She pauses for a moment, to let the logic of this sink in. "Now, I hope, we hope, that this is just a misunderstanding about how many mice there were and who did what. We just have to deal with it objectively. So let's have a look at your notebooks and whatever supporting data you have."

"But what you do mean by supporting data? As I said, the results are just numbers. I counted the number of tumor nodules, wrote it in my lab notebook and calculated the average and so on for the graph."

"Did you take pictures of the mice, showing the tumors?" Tom asks. "That would prove how many animals you examined."

"No, I just dissected the mice and counted the tumors. I guess I should have photographed them, but I didn't think of it at the time."

"That is too bad." Sushma says. "Photographs would certainly have helped. The main issue here seems to be the number of mice you analyzed. So let's focus on that. Is there any other evidence to document that you analyzed six double transgenic mice?"

"I am not sure I would have anything other than the tumor counts."

"OK" Sushma's voice now has an edge of impatience. "Well, how about any evidence that six mice of the right genotype existed when these experiments were initiated."

Tom perks up, "What about the PCRs from the genotyping, Chloe? That should do it. You did those yourself, right? So you should have pictures of the gels or the sequencing data."

"No, I don't think I kept the genotyping data. It was all a bit hectic at the time. We were under a lot of pressure with the revisions of the paper. I focused on getting the results."

Tom tries again. "But Chloe, surely you must have something that can help us out. Maybe you still have the tail DNA preps that you used for genotyping the pups? In the freezer somewhere? Or maybe you have frozen tumor samples, or histology slides from these? We normally do histology as well. Even if you did not use them for the figure, such slides would show how many samples you had. If they are labeled properly."

"I'm not sure. I don't think I kept the DNA samples."

"You have to have something." Tom says, exasperated. "Don't you understand? Otherwise. . . Otherwise, I don't know. We should have a look at your lab notebooks. Hopefully, they can help." Tom looks at Sushma for confirmation.

Sushma shakes her head very slightly. When she speaks again, her tone is more formal. "I do hope the members of this lab understand the importance of keeping good records and preserving data. Apart from everyone being required to attend the ethics course, I expect that you, Tom, have made this clear to your students and postdocs." She looks firmly at Tom, establishes eye contact and keeps it. If he has not been vigilant in ensuring good recordkeeping and so on in his lab, he is also at fault. She is reminding him of that. Tom breaks eye contact, hesitates, then starts to explain. Sushma signals for him to wait. They can discuss this later.

"I suppose you are right." Chloe finally says and adds, with a note of sarcasm, "But it is not my impression that recordkeeping has ever been a major priority in this lab. Not that I have ever heard of." She glances at Tom, but he is looking off to the side. No one says anything for a while. Chloe continues, her voice changed again and now almost pleading. "The point is, I did what I said I did and I have the numbers in my lab book. But I don't know how I am supposed to prove anything beyond than. I really don't." She pauses, getting flustered. "I might have some histology slides in the fridge. Yes, I should have some slides in the fridge." She adds, finally.

"OK, Chloe." Sushma seems to be wrapping up. "Let's take this one step at a time. First, we will go to the lab together. You show me your lab notebooks and we will look at what else you might have. Hopefully we can find the histology slides you mentioned. If so, I hope there are precise records of them in the lab notebook pages as well." She starts to get up. Tom also gets up but she stops him with a subtle hand gesture. "You don't need to come along for this. Chloe can show me." Turning to Chloe, she continues in a tone that remains formal, but also courteous.

"I'll go with you to your bench to look at the items we just discussed. If anyone asks, you are just sharing data with a colleague." She pauses, "But I recommend that you go home afterwards, Chloe. Take some time off. What happens next depends on what we find in the lab. I may have to take some of the materials with me. In any case, I will need a few days to confer with my colleagues."

They all get up, somewhat awkwardly. Sushma says goodbye to Tom with a firm handshake and a promise to get back to him as soon as she can.

Tom remains standing after Chloe and Sushma have left the office. He is clearly rattled. He had expected some resolution of this today. He had certainly not expected to have his own standards or his way of running the lab questioned. Worse, he had no opportunity to reply. He will have to explain to Sushma and to the director's office that he is not lax about recordkeeping and evidence. This is not his fault.

Sushma follows Chloe into the lab. Chloe's expression is hard to read, neutral, almost mask-like, as she walks briskly through the long room. When she reaches her bay, she sees Juan is there. He turns around and says a friendly "Hi, welcome back". Getting no response, he looks at Chloe, puzzled. He notices the unknown woman in the suit trailing her and turns back to the paper on his desk.

Standing at her desk, Chloe picks up one of the lab notebooks from the ordered array, opens it and searches for a page. She finds it, points out some details and hands Sushma the notebook. Sushma asks a couple of short questions and Chloe shakes her head to each of them. Next, they go to the small fridge below Chloe's bench. Chloe opens it and crouches down to look inside. She takes out a stack of slide holders, opens a few of them and shakes her head slightly each time. She replaces the slide holders. For a while, she just sits, still looking into the open fridge. She turns around, looks up at Sushma and shakes her head one more time, very deliberately. Sushma understands and nods. Chloe gets up slowly. Sushma picks up her bag and the lab notebook that Chloe gave her. She shakes Chloe's hand, and leaves the lab quietly.

Chloe pulls the chair out and sits down, staring at the dead computer screen. Her expression moves from frozen to shocked to determined. Juan turns around to chat. Seeing the look on her face, he thinks better of it and returns to his reading again.

She feels numb. The last few weeks, she has been soaring. The world at her feet and so much ahead of her. And now this. It is crazy, unreal. This cannot be happening to her. Envy, that must be what this is triggered by. It must be. But they will not get away with it. They cannot take this from her. She gathers her things and gets up from her desk very briskly. Juan and other lab members nearby steal glances in her direction. At the other end of the lab, Tom opens the sliding door and sticks his head inside. He sees that she is packing up and goes no further. He probably wants to make sure that she goes home as arranged, but is keeping his distance. You weak son of a bitch, she thinks, hiding behind procedures. You didn't even try to help me in there. She stares at him for a moment, hard, before turning toward the other door.

Chloe leaves the lab quickly, without a word to anyone. She needs to get out of here, immediately. Away from these people and their petty jealousies, away from Tom and his cowardice. But she is not beaten. She will fight this. They cannot make her a scapegoat of lax lab procedures. Right now she just needs to get out of here. Then she will see. But she most definitely needs to get the hell out of there.

Once out of the building, Chloe realizes she could have gone to see Martin. But Martin will not want to hear about this, she is sure. He would rather remain safe in his innocent little world upstairs. She cannot focus on him right now. She doesn't even know what to say. They can talk tonight. She starts walking. Pretty quickly, she realizes to her surprise that she is hungry. But she does not want to bump into anyone she knows just now, least of all anyone from the lab. She needs more distance. So she keeps walking, toward downtown. She'll find something there. The walk will help clear her head, she thinks.

–

That evening, Chloe brings along the bottle of Chianti that she bought on her city wandering. It is nothing special, just wine. Martin seems surprised when she hands him the bottle on her way in the door.

"Wine on a Wednesday? What's up?"

"Well, you just got me back. That should be worth celebrating for more than one night, no?"

"Absolutely" He says and pulls her close.

"Food first" she insists, pulling away. She opens the wine and pours a glass for each of them.

"So how was the first day back from the tour? What did Tom say?" He asks, quite naturally. Last night, she talked at length about the interviews and mentioned the likely job offers. She was alight with excitement, even if she was trying to restrain herself. He can see that she is more subdued today, but does not immediately read anything into it. He turns to put the spaghetti in the pot.

Chloe has been walking most of the day and has not talked to anyone. But Martin will have to be told.

"Well, you won't believe what happened to me today. My big welcome back. I was ambushed by a bunch of backstabbers." She says.

"What do you mean, ambushed?"

"Exactly that. They had planned it, gotten it all ready and ambushed me with no warning. Tom is such a coward, he just went along." Martin stops the dinner preparations and turns around to face her.

"Wait a minute, what are you talking about? And who are they? You need to give me some more information here."

"OK, so someone in the lab, some nosy, jealous idiot, told Tom that they thought there might be something questionable in my Nature paper." She puts down her glass and looks at it, pausing, then looks back up at Martin. "Someone in cahoots with that lazy mouse technician who has been with Tom forever."

"Andy?" Martin asks.

"Yes, Andy. Well, they decided that since Andy does not have correct records of the last mouse experiment that I did—that I did, mind you, not Andy—then there must be something wrong with my data."

"What? That's stupid. Surely that would be his mistake, not yours."

"You'd think so, wouldn't you? But instead of asking me about it, the two of them go to Tom and tell him they are terribly worried that there might something wrong with my published work. Can you believe it? They question my work, because Andy keeps sloppy records? I don't know for sure who the other person is, who started this or why. But I know that Andy doesn't like me. You remember how upset I was with him last summer?" Martin nods. "Anyway. So what does Tom do? Does he wait until I am back and he can have a decent conversation with me? No. He gets so afraid of the whole whistleblower, misconduct bullshit that he brings in a lawyer immediately, from the director's office or something. And then he has her talk to me instead of doing it himself."

"What? I can't believe it. That's crazy. Tom seems like a decent guy. How could he fall for this?"

"Completely incredible, isn't it? After everything I put into that paper. Tom was perfectly happy to put himself as senior author on it a few months ago."

"He must be terrified because of what happened to that Bolinger guy." Martin tries, tentatively. "The guy who stood up for his people, for their results, when everyone else was screaming misconduct. He was forced out, you know?"

Chloe glares at him.

"I don't give a shit about Bolinger. Whatever he did or didn't do is completely irrelevant. I don't even want to think about Tom. He is not the victim here. I am. I can't believe that he just hung me out to dry like that. He wouldn't even listen to my side of things first. He just assumed I was in the wrong." The indignation in Chloe's voice is strong, stoked by a long day of solitary walking and thinking.

"So what happened? When you talked to the lawyer?"

"She seemed all nice, like she was on my side and everything. And of course I told her what really happened, that everything was correct."

"So you set it right?"

"Yes. I did. But then she started going on about whether I could prove every little detail or not. Did I have the hard evidence or whatever, the raw data, to prove that my numbers were correct? She wanted me to show her immediately. And I didn't have everything. So then she and Tom got all sanctimonious about keeping good records. As if Tom cares one bit about all that when we write papers. You remember what it was like last summer when we were doing the revisions? It was push, push, push. Get those results, preferably by yesterday. There was not one peep about keeping long-term records of genotyping results and what have you. There never has been. He saw the results and didn't ask for anything more. We put the results in the paper. The paper was published. And when there is just a hint of trouble, he suddenly pretends that yes, of course, he always tells people to keep records of every last miniscule detail. If Chloe didn't do that, well, that's her mistake."

"That must have been terrible for you."

Martin gives her a friendly rub on the upper arm. She seems too angry for him to risk a real hug.

"And on top of that, the histology slides are missing. They are gone."

"Gone?"

"Yes, gone." She says, now more in control, but still obviously incensed. "Someone must have taken them from my fridge and thrown them out."

"Thrown them out? That's just insane."

"I know. It's all insane. But I was gone for almost a month. Anyone could have taken them. Whoever is accusing me must have done it."

"But surely no one would go as far as taking someone else's slides? I mean ..." Martin stops, realizing his voice must reveal more than a trace of skepticism. She gives him a quick, hard look.

"But that's what must have happened. Whoever did this was making sure I couldn't prove anything."

"Come on, Chloe. Surely you don't really believe that?"

"So you think that scientists are not capable of doing things like this? You can't be that naïve, Martin. Scientists are just people. There is no shortage of jealousy and envy and what-not to move people along."

"Well, yes, but—but Tom doesn't seem like a bad guy. He is probably just worried they will come after him if he is not seen—" Martin makes quotation marks with his fingers "- to be 'doing the right thing'. You know. Are you sure those slides were not just mislaid? Maybe if you look around. . ."

"Of course I am sure." She gives him a look that allows no further questions. "No, no way. They cannot just accuse me of something without any proof whatsoever. That is just plain wrong. I thought you were presumed innocent, not presumed guilty, around here. I shouldn't have to prove I'm innocent." She takes a sip of her wine and adds, with obvious bitterness "So for now, I have a few days of leisure. While the lawyer, who is also a failed scientist, by the way, examines my lab notebook."

She keeps looking at her glass. She only looks up when Martin turns around to check on his spaghetti. It must be overcooked by now.

"So how was your day?" She finally asks his back, with a forced smile.

Chapter 11

"Someone took them," Chloe says in a firm voice. "Someone took the slides from my fridge." She is all rigid determination, sitting in the chair opposite Tom's and looking him straight in the eyes.

"Took them? Are you sure?" Tom asks, momentarily stunned by this denial and accusation in one.

It is 2 days after the interview with Sushma. He called Chloe and asked her to come in to talk about how they should proceed. After 2 days of contemplation, he imagined that she would come to him with an apology, possibly a tearful admission of a mistake made under stress. Maybe without Sushma there, such an admission would be easier. Or even better, she might have remembered where the tail preps were or found some other evidence to support her version of events. But no evidence has materialized. And Chloe seems more steely than repentant. She is keeping her emotions under control today.

"The histology slides had sections from the MMTV-Myc plus Ras tumor samples, the inhibitor experiment we discussed. The slides would prove that I analyzed six mice, exactly as I stated in the paper. But now they are gone. I mentioned the slides when we had our discussion with the university lawyer or whatever she was."

"Dr. Sushma Nayar, from the office of scientific integrity. And yes, she is also a lawyer." Tom says, thinking how unlike Chloe to be disrespectful in this way.

"Whatever. I told her that there would be histology slides in the fridge to back up the numbers."

"I remember you saying that you might have such slides, but also that you were not sure. And now you are sure that you had them?"

"Yes, I am sure. It was just that Wednesday was so. . . Well, I was rattled and I wasn't thinking straight. It seemed like—and I have to say this—like an ambush with the two of you. I was in shock. Now that I have had a chance to think it through, I am positive I had the samples. They should have been there. I looked for them Wednesday, when we went back to my bench. But someone must have removed one of the slide holders while I was away. The fridge is not locked or anything."

"So you are saying that someone went into the fridge under your bench, found these samples and then deliberately got rid of them?"

"Yes, probably to discredit me. So I am pretty sure it must be the person who is accusing me."

"But wait a minute, Chloe, this is ridiculous. No one would do that."

"It is no more ridiculous than accusing me of making up results. I would never do such a thing." Chloe glares at him.

"But what about Andy's records?" Tom protests "What about the missing genotyping data? There should be some other evidence around of your having worked with those mice. You seem to be missing all the raw data for that one experiment."

"I explained that already. Things were very hectic when we were finishing up the paper. You know that. We did not discuss what exactly had to be kept, just the result."

"But you should still have some evidence that these mice existed. There should be a record of them, in the lab or in the mouse house."

"It was half a year ago, things get lost. I have the numbers in my lab book."

Chloe keeps her eyes on Tom and continues evenly. "Look, you did not question anything when I came to you with the results originally. Why didn't you ask to see genotyping data then?"

"I trust you guys when you show me the numbers."

"And now you don't trust me? That seems awfully convenient. There has never been a great deal of emphasis on meticulous record-keeping in this lab." She adds, almost sweetly, "Or maybe I missed that session?"

"Chloe, you are an experienced researcher. You know these things. You attended the ethics course like everyone else, where everything is spelled out."

"Only in general terms, not what needs to be kept in what situations."

"Come on Chloe, don't play at being naïve. You know perfectly well that you need to have good records and keep original data, especially for a key experiment in a publication."

"But genotyping data? Those are not results. They are just a form of identification. We normally never look at genotyping data after an experiment has been started. They are never presented and you never said they should be kept on record. In retrospect, it is obvious that it would have been good for me to have kept them. But how was I to know all this would happen? The actual result of the experiment was in the tumor burden at the endpoint. I counted the tumors and the numbers are in the lab notebook. The tumors are also documented in the histology slides. Well, they were. But someone took them."

"But what about photographs of the mice? In lab meetings, we use pictures of dissected mice to show the appearance of tumor nodules all the time." Tom says, in a tired voice. They are going in circles.

"We talked about that already. We don't always photograph the dissected animals. I didn't, this time. I counted the tumors for each animal. You have seen the results."

Tom shakes his head slightly.

"Look, Chloe. Here is the situation. Sushma has spoken with her colleagues and gotten back to me. This is why I asked you to come in." He looks at her. "Basically, they are not convinced that the whistleblower's concerns can be dismissed easily. So they want a full investigation."

"A full investigation?"

"Yes, of possible misconduct."

"Misconduct?" Chloe says, her voice less steady. "Tom, you cannot mean that. You know that once someone gets that label, misconduct, science is over. I am a good scientist. You know that. I do not deserve this. There is no misconduct." The word misconduct seems to dig in further with each repetition. "You have to believe me. You have to." The switch from cool disdain to frazzled pleading is surprisingly sudden. Then she hardens again. "And especially not based on malicious accusations. The real misconduct here is someone stealing my slides and making false accusations."

"At this stage, it is just an investigation, looking closer to find out what happened. There is no verdict." Tom says, with deliberate calm. The truth is that even a hint of misconduct can do a lot of harm. But acknowledging this will not help in the present situation.

"Anyway, we have no choice in the matter. They will bring in external observers and they want your computer, your lab notebooks and full access to all materials in the lab and in the mouse house."

"I will have to tell them about the missing slides."

"Well, I'll be dealing with them, with help from Lucy. I will tell them what you just told me."

Chloe's expression is back to one of disbelief.

"Are you completely forgetting, is everyone forgetting, the huge effort I put into this paper? I have tons of raw data, lots of evidence for the findings in the paper. It is a solid paper. It is not make-believe."

"Well, yes, we also have to talk about the paper. I know you put a lot of work into it. But if, and I really hope that this will not be the case, but if these results are not properly supported, we may have to retract the paper."

"No. No way. We cannot retract the paper. That would be totally wrong. Apart from how much it would hurt me and hurt you, it would just be plain wrong. For scientific reasons." She is more forceful again, arguing instead of pleading. "The data in the paper are correct. It is important work, and you know that. That someone has stolen what I need to prove one last little point cannot mean that a whole body of work gets trashed. That's just not right."

"I hate the idea as much as you do, but it may be necessary, given the uncertainties we now face."

"Hate the idea much as I do? I really doubt that." Her bitterness is raw, unchecked. "To me, this paper is everything. It represents 4 years of my life. It is the foundation for my career. It is not the same for you. It is just another feather in your cap." Tom tries to respond, but she silences him with a hard look. "Anyway, as you said, these crazy accusations are being investigated. It is not decided. It would be wrong to retract a paper just because there is some unproven objection out there. I have explained about the mice and about the slides. The numbers are in my lab book. There is nothing wrong with the data in the paper." They are both quiet for a moment.

"Look" Chloe continues. "I have a solution. I can redo the experiment that has been questioned. I can repeat it as much as you want, so you can see that the original data are correct. That way, the data in the paper will be confirmed."

She may have a point, Tom thinks. A hasty retraction might not be the best way forward. He should make sure it is the truth that stands in the end. Chloe's suggestion is tempting, but his hands are tied.

"I'm afraid that's not possible." He finally responds.

"Why not? It makes perfect sense. Surely the most important thing is whether we have published the correct result. Finding out the full truth, isn't that what this is all about?"

"Well, yes. But I cannot let you do it. While you are being investigated, you cannot be in the lab. You will have to be on leave." Chloe looks unconvinced. "Sorry, those are the rules," he continues.

"And what about my accuser? Surely this person should be on leave as well? I mean, removing someone else's samples is worse than not having a few genotyping results handy when someone asks for them."

"You mean the whistleblower? No, that is not the same situation. You cannot just claim that someone removed your samples and have that person be reprimanded without any evidence whatsoever. There is no proof the whistleblower has done anything wrong."

"But I am being accused without evidence—" Chloe says heatedly, "—without proof. I am the victim here."

"Chloe, you must understand that it is not the same thing. Unfair as it may seem, you, or we, have to be able to back up the claims in the paper. You seem unable to produce any physical evidence to support Figure 7F and your version of events. No photos of the mice, no leftover DNA from the genotyping, no tumor samples, no slides. And there are the inconsistencies with Andy's records. Noticing an inconsistency that calls published data into question, as the whistleblower has done, is not analogous to accusing someone of stealing. Your situation and that of the whistleblower are not comparable." Tom also knows that he cannot be seen to be punishing a whistleblower in his lab. No, he needs to treat Karen and Andy very carefully or this could get much worse for him.

"I have to ask you to stay away from the lab while this is being cleared up. Take some time off. Work on your grant applications or whatever. But at home. We will keep this completely confidential while the investigation is running, and it will remain confidential unless they find evidence of wrong-doing. I'm sorry, but those are the rules."

"And how long will that be? How long am I banned for?"

"They said it could take a couple of months."

"A couple of months? But that's forever. What about my job applications?" Chloe has realized that the outcome of her job search is likely to be the most precarious part of this situation. Yes, they were extremely positive, but they have not formally offered her a job, not yet. Any whiff of misconduct and it will be over. "You know what would happen if any of these places got word of the investigation, don't you? They would drop me like a hot potato. Obviously. They wouldn't want to take any chances. I would be finished, judged on rumor alone." She sighs. "So it is really important that you keep this confidential. I wouldn't like to think of what might happen if someone . . . well you know. It is only reasonable, I think."

Yet another change of demeanor. Bargaining? Tom isn't sure how to deal with this. He opts for a straight, formal answer.

"Of course we will keep this in the strictest confidence. The rules are designed to protect everyone involved until the investigation is over."

He is having a hard time working out what she is thinking. It must be obvious to her that it is not in his interest to broadcast that a postdoc in his lab is being investigated for misconduct. She has not been found guilty of anything, at this point. Maybe she thinks he has given up on her already. Well, the missing slides story is just a little too convenient. A good offense. . . He wonders how the investigators will deal with that. He shouldn't be too quick to dismiss her explanation, though. She has earned at least some benefit of the doubt. Anyway, that will not be his call, it seems. His job is to decide what to do about the paper. He is the senior author, after all.

During these musings, Chloe has gotten up from her chair and moved toward the office door. Tom gets up as well.

"I'll let you know as soon as I know more." He says "And, of course, the investigators will want to talk to you as well. They will contact you directly, I think. There may also be some practical questions about where to find things. Well, let's just see."

"Everything is in the lab. Or it should be."

"So is there anything you need from your desk or bench? Anything that you need to take care of in the lab before your leave?"

"I've been away for almost a month. A couple more shouldn't matter too much, right?" A note of sarcasm has returned to her voice. "But no, there is nothing that I need from the lab."

"OK, good, I'll walk you out then."

Chloe sends him a quick look—walking her out of the building?

"I mean I'll walk you to the elevator." Tom nods to the left, the direction of the back elevators. "I was going in that direction anyway."

Of course, she thinks, the corridor going by the lab with its wall of glass is very exposed. He is quietly suggesting they use the exit direction that passes through the non-transparent office section instead. He is being considerate. They walk along in silence. As she gets into the elevator, habits of a lifetime draws forth a weak smile from Tom and he mutters. "I'll be in touch."

Afterwards, Tom retreats to his office and sits down without waking the computer. He feels drained, although it is not even lunchtime. Drained, but maybe not done for. It has occured to him that there may be a way for him to mitigate the situation with respect to the paper. Independent confirmation. They should repeat some of the experiments in the paper. No, better yet, repeat all experiments involving Jmjd10 and all experiments with the inhibitor. The screen was a major part of Chloe's work, but there is no reason to redo that aspect of the paper. The screen just led them to Jmjd10. Realizing this division, the job seems manageable. He hopes everything in the paper is correct, but this way he will know for sure. It should not take too long to do. Most of it is cell culture and the reagents have been made already. For the mouse experiments, they have what is needed in house. Luckily. So it should only take a couple of months, maybe three. Then he can determine whether a retraction is needed. Maybe a correction will do. Maybe that won't even be needed, if all the results are in order. Even a correction is by far preferable to a retraction. Retractions attract way too much negative attention. And it's the right thing to do. Why retract a result—even worse, a whole paper—if it is correct? Who would that benefit? The scientific community needs to know what the correct result is, what conclusions they can believe in. Every little complicated step on the road to that knowledge does not need to be broadcast. It never is. Stuart will no doubt tell him that it is a bit more complicated than that. But Tom is convinced this is the right way to go. He wants to know what the scientific truth is, the correct result, and he will find out. Naturally, Chloe cannot be involved. It has to be another postdoc from the lab. Well, more than one if they are to do this quickly. Chloe will no doubt be angry at the thought of someone else working on "her" project. But, he thinks, didn't she sort of suggest this herself? She also wants to avoid a retraction, if at all possible. So he can talk her around to his point of view, if required. For now, he does not have to explain himself to her or to anyone else. He just needs to move forward.

He has Deidre set up separate appointments with Lucy, Hiroshi, Juan, Yuqi and Karen. This afternoon. He will make a plan over lunch and get everything going by the end of the day. It feels good to be taking action.

"But why me? Can't someone else do it? I need to get my project to publication. You know the time pressure that I'm under." Karen is visibly uncomfortable. "And I really don't want to be involved in Chloe's story. I just noticed something was odd and I told you about it. And I talked to the Indian woman from the university, as you asked. But that's it."

Tom is getting impatient. "Look, Karen, you started the ball rolling. You, of all people, should be willing to help find the truth here." He looks straight at her, challenging her. "You do want the truth, don't you?"

"Of course I do" she answers meekly. "But maybe it would look bad? I mean, me involved in the re-checking the paper? Given that I was the one to point out the problem? Maybe it would be better if someone else did it, someone completely neutral?" she asks, faintly hopeful.

"Don't worry, you won't be the only one involved. I have asked other postdocs to help out as well. It is a quite large piece of work, after all. We need to get this going immediately and we need to get it done fast. So we, as a lab, can get this issue settled, once and for all. This is why I need you and the others to pitch in with time and effort now."

"OK, I understand, I suppose." Karen sinks back in the chair and sighs. "So what is it you want me to do?"

"Well, you and Andy have set up more matings to get MMTV-Ras, MMTV-Myc double transgenics, right?"

"Yes, for my experiments, as we discussed at the group meeting. I really need those mice." Karen finds herself pleading. Surely, he will not punish her for speaking out by making it impossible to save her project?

"Right, you will want to do that as well, of course." Tom says. "But there will be plenty of mice coming. Do the genotyping carefully and document it. I want you to use the first six double transgenic pups to repeat the inhibitor experiment in Figure 7F. Set up paired drugged and control mice. Also, when the experiment is running, document everything carefully. Have Andy sign off on it. Take pictures whenever possible and keep a record of them in a hardbound lab notebook. Keep the tumor samples at the end, ready for histology. You get the idea?" He knows this may sound a bit excessive. But there can be no doubts this time around.

"Yes, I get the idea. I suppose I understand why you want this, but—", Karen hesitates, "—what about my experiments? Isn't my project important as well?"

"As I said, this is just for the first six double transgenic mice. The rest of them are yours to do with as you please. Just set up more matings. There are plenty of MMTV-myc mice." He says, exasperated. "A delay of a few weeks will not make a big difference in your situation."

She does not need Tom reminding her of "her situation". She was feeling just the tiniest bit uplifted after the group meeting and her discussion with Tom. She was looking forward to some progress. Now this. The extra work will take time. And it will delay her getting the mice she needs for her project. A small delay, yes, but still. Her work is being sidelined for Chloe's project, for their precious Nature paper. Repeating the contents of a published paper, what a waste of time, she thinks. She remains quiet, however.

"I also want you to do the other inhibitor studies in Figure 7. Look carefully at the corresponding part of the text and at the figure. Every last statement and figure item has to be re-tested. You are the person in the lab most familiar with the 3D cultures. That's why I'd like you to do it. I'm confident you can do the experiments properly. Why don't you write up a detailed plan and show it to me tomorrow? We can then check that everything is accounted for."

Tom looks determined, not to be contradicted. She isn't going to. She knows she cannot afford to antagonize him. She nods.

"Here, as well, everything is to be documented carefully. All manipulations done and all results obtained have to be noted in a lab notebook, including pictures where possible. I want the raw data accessible and interpretable. Even if something looks wrong and you are tempted to throw out a plate or whatever. Everything must be fully documented. Lucy or I will sign off on the relevant lab book pages every week. All those involved in the 'repeat project' will meet on a weekly basis anyway, to share results."

She nods again, resigned. So many experiments to do, and minute documentation of every step. It will be a huge amount of extra work. One small consolation is that he is not just targeting her. Others have had similar conversations with Tom and have been burdened with this extra work 'for the lab'. Still, she wishes he would spare her, given her recent tribulations.

Somewhere in the back of her mind, she has started to accept the logic. It has to be done. Not for Chloe, but for Tom and for the lab. The reputation of the lab matters to her. It is a lose–lose situation, however. The best the lab can hope for is that the data are correct and do not require retraction. Which means their extra work will have been quietly wasted. The worst result is a retraction. And who would want to be part of that? The actual science reported in Chloe's paper, maybe this should matter to her as well. But this feels moot now. The results have already been published. Enough, she admonishes herself, stop whining and get on with it. She starts to get up from her chair.

"Oh, and get the inhibitor stock solution as well as anything else you need from Lucy. She will have everything. I've already told her to be ready to help those of you on this mission."

Mission. Strange word to use in this context, she thinks. What kind of mission are they getting into here? But she keeps this to herself as well. She simply nods again and heads out the door.

What a day. Tom glances at the clock on his phone. Almost 6 P.M., he should be getting home. But this crazy day requires a bit of digestion. Since he finished talking to Chloe this morning, he has been fully occupied putting his plan into action. He has talked to each of the postdocs that he selected, one at a time. Lucy as well, of course, but she was easy. She is used to being given instructions. The postdocs are not, and the conversations were not all smooth. He usually never tells them directly what to do. He gives guidance, encouragement and a push when needed. This is their covenant. They are all excellent postdocs by any standards. Each of them got an independent fellowship for at least part of their time in the lab, so it is natural that they feel they should be working exclusively on their own projects. And under normal circumstances they would be. But they have to understand that this is an extraordinary situation. He gave them each the spiel about how sorting this out was important for the scientific community and for the reputation of the lab. Forget that this is Chloe's paper, he said, think about the science and the lab. They eventually came around and agreed to do their bit. The surprise was how reluctant Karen was. She was sullen, almost, and seemed to feel that it was a terrible imposition on her time. He almost lost his cool with her. She started this, for God's take. She should be the first to help out. And hadn't he, just last week, told her he would find support for her when her fellowship runs out? Shouldn't she be just a little bit grateful? Sometimes, people's inability to see beyond their own self-interest surprises him.

Unavoidably, his musings lead him to the first conversation of the day, with Chloe. He still cannot make himself believe her counter-accusation about stolen slides. Discarding someone else's samples is a serious offense, if done with malicious intent. Despite his irritation with Karen just now, despite his impression that she may be jealous of Chloe, he just can't see her doing something so strange. He considers the precautions he has taken with respect to Karen sensible, even necessary. But he does not believe she removed Chloe's samples. Envious, perhaps. But she would not go that far. And then there is

Andy. Andy has no reason to lie. At worst, he could have missed something or forgotten. Not likely, but also not out of the question. That leaves Chloe.

Could she really have fudged the data? She knows the consequences of being found out would be devastating. Her life in science would be over, finished. No matter how smart and otherwise accomplished she might be. If she is found guilty, it is over. But even a hint of misconduct could be enough to derail her career. Chloe was right about that. So the idea of her doing something like this is crazy, really. But oddly, he finds that he is able to imagine it. The pressure of getting an important paper accepted is huge. Cutting a corner in the last experiment needed for a major paper is completely and absolutely wrong. But, but. He thinks of last summer. Everything else was in place. The reviewers were positive but unreasonably picky and demanding. That last experiment seemed completely superfluous. He even remembers saying something like that out loud, in a flash of impatience. So the temptation is understandable. He does not condone it, of course not, absolutely not. But, it does make it easier to imagine that Chloe is at fault. Still, it puzzles him that she has not given in. She maintains her innocence. And the outlandish counter-accusation, that surprised him. It makes him doubt his gut feeling. However improbable it seems, maybe she is telling the truth. Or, maybe she just doesn't give up a fight, not ever. That would certainly fit with her personality, as far as he can tell. She has pluck and drive. He almost wants her to win. Then he checks himself. It is not a game for someone to win. It is about figuring out the truth, in more ways than one. And today he has done all that he can to make that happen. Still, he shouldn't kid himself. If the investigation finds misconduct on Chloe's part, he will take some of the fallout as well. It will look like he has been too lax in his supervision, or even worse, encouraged shortcuts. A bit of schadenfreude is to be expected. It may get ugly, but as long as he acts swiftly and does not appear to be concealing anything, it won't get too serious. Stuart was right, get in front of it from the start. His job is safe, his reputation only slightly exposed. Then there is the paper. He would hate to lose it, to be forced to retract it. Apart from the embarrassment, it is a great story. And, if he is honest with himself, he needs it. It has been a couple of years since his last major paper. Too many years without a high-profile paper and your reputation starts to fade. You become a has-been so very easily. Anyway, it seems wrong to retract the paper before he knows the full truth. The truth about what Chloe did but also the true results. Even if she did not do the final experiment properly, the result is most likely still correct. He feels sure it must be, given everything else in the paper. The opposite, the worst-case scenario, has occurred to him as well. Other results in the paper could be faulty or manipulated, not just Figure 7F. He shudders. No. Stop Get the facts first, and then decide.

It has been a long day, he thinks, finally rising from his chair. The next couple of months will be trying, for all of them.

Chapter 12

A downtown business district of a futuristic city. Row upon row of high-rises. All made of glass. And all with bright red floors. You just need a bit of imagination.

Karen almost smiles at the imagined likeness. Then she sighs. The stacks seem endless. The plastic cubicles are each neatly labeled, but otherwise identical. And they are not for display. They all require her attention, each and every one of them. So many samples, so much to do. It is enough to make even the most steadfast trooper weary, she thinks. She transfers one of the stacks, dish by dish, from the incubator to the tissue culture hood. I'll get to you guys later, she thinks and closes the door on her colorful cityscape. Sitting back down, she focuses on the vials lined up in front. She picks up a labeled vial, pipets a few microliters into one of the twenty tubes in front of her. She does the same with the next vial. Soon the mixtures will be added, each to one of the labeled dishes. Repeating Chloe's experiments is tedious. But she must keep her focus, pay attention to every tube, every step performed. The plan for today's experiments is pasted into the lab notebook on the stool next to her. Every detail of everything she does will also be recorded in the book: How much, exactly, from which tube, how many minutes incubation, when it is completed. It requires constant attention, but not much thought. She is unable to concentrate on anything else, though. Very irritating. If she is to sit here for hours on end, why can't she at least think some useful thoughts? The noise is also starting is get under her skin, the relentless drone of the hood. Once you start noticing, it becomes hard to ignore.

Like most other mornings, Karen has been at her spot for several hours. Yuqi is in the hood next to her. It is hard to tell whether she is as bothered as Karen is. Karen looks over for a moment, but Yuqi does not seem to notice. She is working with a large set of multi-well plates of her own. Karen knows that Yuqi also has a set of preprogrammed experiments to go through, her part of the repeat effort. Like Karen, she keeps her own project going alongside the work for Tom and puts in the extra hours needed. Karen is impatient to get on to her real work. This is tiresome and very boring.

Stop whining, she thinks to herself, not for the first time. It is just for another month. Why is she letting it get to her so much? The actual work is not so different from what she usually does. But, it is. It is the difference between being a postdoc and a technician. Doing experiments because you are so keen to know the answer that you cannot wait is one thing. Doing it because someone says it needs to be done is completely different. If she were doing this for her own project, she would be perfectly content. She remembers trying to explain this to Ashok a couple of months ago. It probably made little sense to him. He is a thoughtful person, but not an experimentalist. To him, an experiment is an abstract entity, the step between an idea and the result. To her, an experiment is also the process. A satisfying, engaging process or a stupefyingly tedious one. Everyone knows that lab work is full of repetitive tasks you've done a thousand times before. Miniprep DNA, run a gel, bla, bla. You could practically do it in your sleep. But if it is your idea and design, your baby, everything changes. She smiles to herself recalling the baby analogy from her time as a PhD student. She heard various old friends, now new parents, talk about their babies. Bright eyed and amazed, they marveled at their own creation, so very special. While other people's babies remained perfectly predictable and boring. Maybe it is related, maybe not. Experiments do not have personalities, of course. But an experiment that you designed and where you care about the outcome is your baby. A similar one, preprogrammed by someone else, is just work. This is dead obvious.

Repeating someone else's already published experiments is proving to be the worst possible scenario. Boring, yet demanding. Although Karen knows she shouldn't, she resents being on the repeat team. She imagines the three other postdocs on the team resent it even more. They must be pissed off at having to clean up Chloe's mess, whether they admit it or not. Sure, Tom has promised they will be given credit in some way, but still. They all have their own projects and this cuts into their time. The competition isn't sleeping or having a holiday.

At one level, Karen knows she did the right thing, bringing her concerns to Tom. But she doubts anyone would thank her, if they knew she started this. She hopes they don't know. It is hard to tell. Sometimes she feels that they see right through her. So she is uncomfortable around her lab mates, more often than not. But the feeling of unease is not hers alone. The atmosphere in the lab has changed since all this started. No one could miss Chloe's long absence and everyone heard about the repeat effort. So Tom knew he had to explain. He did a fair job of it, she has to admit. At the beginning of a group meeting, he announced very calmly that it had come to his attention that some results in their recent Nature paper might be problematic. He did not specify which results. He explained that he had put together a small team to repeat the experiments. It had to be done independently, so Chloe was not involved. He emphasized the importance of not leaping to conclusions. "We don't know exactly what happened." he said. "The job of the team is to check everything. And I ask all of you to keep an open mind, both personally and scientifically.

OK?" He then reminded them sternly of the need for confidentiality. "We have to protect the people involved, until we know the truth. So you are not to discuss this with anyone outside the lab until we are done. Absolutely no idle chitchat with anyone, is that clear?"

Ever since this talking-to, interactions in the lab have been strained. Even Vikram has been subdued.

Thinking back on Tom's words 'come to his attention', Karen is sure that everyone must have guessed that there was a whistleblower. They may even have guessed that it was her. At her most recent group meeting, they discussed the mice that were used in Chloe's paper. Someone might put two and two together. But at least she was not officially exposed, as it were, by Tom. He went through his spiel about the scientific truth being essential, which was why they had to redo the experiments. He is not wrong, perhaps, but it is unusual to do this and it is a big effort. None of the postdocs in the lab are naïve, far from it. They must sense the strain between Tom's self-interest and his 'finding the truth for science' line. Whatever, it has to be done, 'for the lab'. And the repeat team all want to have it over with, so they can move on to other things. But for now, the days are long and the atmosphere tense.

The last hour of the morning finds Karen in a very different mood. She is setting up the delicate 3D cultures with her specially designed cells. She is back at the cutting experiments she started a few months ago, but with one big difference. Now, it works. At the end of every afternoon, she is on the microscope in the Küstner lab, doing the manipulations and setting up recordings overnight. Crazy, she thinks, all this other stuff is going on and finally her intractable key experiment starts working. The trick was to get the optimal adjustment for target depth and placement, plus a lower laser-power and a bit more patience. Now she can do the manipulations she wants and get minimal collateral damage. Greg, the engineer, helped her figure it out. Once he realized what she was trying to do, he got interested in the problem and seemed to take solving it as his own personal challenge. And together, they cracked it. Since then, Karen has been spending all the time she can, which means all the microscope time she can get, on this. She's been getting reliable results for 2 weeks now. And it looks promising. As she had hoped, disrupting the nanotube connectors appears to have a specific effect. But it is even more interesting than she had expected—cutting the nanotubes affects both the sending and the receiving cells. It is really neat. Still, she knows that this will not make a top paper. The Science paper, published almost 2 months ago now, stole her thunder. They got there first. That cannot be changed. But she has a cool result and she will have a nice story, enough for a good paper. Where she can publish the paper depends on the tumor experiment, her final experiment, as well. But she no longer sees the project as wasted effort, a dead end. It will be a paper she can be proud of. Of course it still hurts to think that the official discovery, and a Science paper, could so easily have been hers. If she had only published her initial observations without trying to push for the cutting experiment, she would have gotten the credit. If only. . . But she has to get over that. She is getting good results. And despite constantly reminding her of the missed opportunity, working on her project makes it easier to deal with the drudgery of the repeat work. The days just get extra long. A five-to-nine job, is the lab joke.

Only later in the afternoon will the microscope downstairs be free for her to use, so she will have to start after the 4 P.M. meeting. Anyway, she likes to keep this part of the work as a treat for late in the day. The tranquility of the microscope room is then something to look forward to. After lunch, she will be in the mouse facility, checking on her mice and further testing imaging procedures. She thinks she will be able to detect her cell markers using the skin flap set-up that Vikram suggested. At least it worked with the first controls she tried. Whether it works with the fast-developing tumors, well, she'll just have to find out. Maybe she can start this evening. No, not possible, not with the repeat team meeting, the second floor microscope work and so on. She will have to start tomorrow. But from now on, every double transgenic pup that is weaned can be used for her experiments. She has completed setting up the repeat of Chloe's mouse experiment. Finally. It took quite some time. Some of the pregnant females produced only one pup of the right genotype, some even none. She is more convinced than ever that Chloe could not have done the final experiment the way she said she did. There is no way she could have gotten enough double transgenic offspring from one or two females. No way. But, whatever, they will know the result in 4 or 5 weeks, when the tumors have had time to grow, or not. In the meantime, there is routine monitoring to do, and cell culture experiments with the inhibitor to wrap up. Plus her own 3D culture experiments and the skin flap imaging experiments to get started. Super-busy time, but also exciting.

Between the microscope room, the tissue culture room, the mouse colony and the procedure room, Karen stays busy. Even with the long hours, she spends limited time in the main lab. Despite her irritation over the imposed extra work, she has to admit that this suits her. This way, she can largely avoid the danger zone of Chloe's bench and desk. When she does pass by, the feeling still hits her, the wave of shame and regret, especially if Juan is nearby. The feeling is not as sharp as before, but it still exerts a dull pressure, strangely complicated by Chloe's long absence. Computer and lab books are missing, but otherwise the place is untouched. It is eerie. It reaches out for her, reminds her. She had no business there. She should have let it all be. At the same time, the image of Chloe herself has mercifully also dimmed, with time passed and the newspaper clipping long gone.

Some days the Chloe issues are impossible to ignore, even without reminders from Chloe's lab space. Today is one of those days. One day a week, the repeat team, Yuqi, Hiroshi, Juan, Lucy, Tom and her, meet up to exchange information. Juan being there makes it extra difficult. And the whole endeavor is of course about Chloe's results and whether or not they can reproduce them. So she is on everyone's mind, even if they don't talk about her directly. These meetings are quite unlike regular group meetings. Normally, no one knows what the answers are supposed to be. Someone presents new results. They report what they have observed and what they think it means. If it is a good group meeting, the others weigh in with possible artifacts that need excluding, alternate explanations and hypotheses and maybe ideas about additional implications. It is all unchartered territory. This is completely different. The repeat team has a specific job and they are doing it very, very carefully. But they are not learning anything new. Karen still feels a nagging unease about the point of the "mission". So what if the published results turn out to be correct? That does not resolve whether Chloe took a shortcut. But would it mean the paper should stand, as is, or not? They have not discussed this explicitly. Tom has simply stressed the need to confirm the results, or refute them. To be fair, he seems open to both possibilities. Karen tries to focus on this, on the necessity of it. The repeat team meetings are subdued but efficient. They get the reporting done as quickly as possible. Karen wills herself to be calm and professional. She says exactly what she needs to say but never anything more.

She and Juan are in forced contact at these meetings. Mostly, he seems to interact normally with her. But sometimes, she is less sure. He has a way of making eye contact that can be unnerving. Not staring, but firm and challenging. What is he thinking? About the whistleblowing? Or does he suspect her of something more, given their early morning encounter? He and Chloe were friends, were they not? Maybe he was always like this, inspecting the world closely. She does not remember, or didn't notice. Maybe it is her imagination. She really prefers just to be working hard, by herself. But on Wednesdays, repeat team meetings have to be endured. One week at a time.

It is four forty-five and they are about halfway through. Hiroshi and Juan have presented their results for the week. Hiroshi has been looking at various effects of Jmjd10 knockdown. Juan has started chemical tumor induction experiment on the Jmjd10 mutant mice and the control mice. Everything is working as expected. The numbers aren't exactly the same as in Chloe's paper, but close, as is normal for biological repeats. More results will be added in the next couple of weeks, but so far, everything fits with what is reported in the paper. Tom is sitting at the end of the table and takes notes as they go along. He has a much-annotated copy of the paper next to his notebook and checks it frequently. During each of the short presentations, he carefully confirms details and probes for possible problems. He appears content. So far, so good. Both Juan and Hiroshi confirm that their remaining experiments are on track and should be finished in 2 or 3 weeks, as planned.

It is now Karen's turn. She looks down at her notes, unnecessarily. When she speaks, she focuses on the projected slides, occasionally looking at Tom. All her experiments to test the Jmjd10 inhibitor have been initiated, including the mouse experiments. The pups are now all weaned. They will go 5 weeks with or without the inhibitor in the drinking water before the evaluation. This is the slowest of the experiments the team is doing, so Tom is concerned about possible delays. But she is on track. She tells them how many mated females were used to get the twelve double transgenic pups, a comment meant just for Tom. She notices that he notices. Originally, the plan was to use six pups, but Tom made her use twelve "just in case". He asked her to test both MMTV-Myc and MMTV-Ras as the female parent, six each way, to be sure that the outcome is not affected by which oncogene comes from the mother and which from the father. But Andy did the cross only one way. Maybe Chloe did it the other way as well? Karen is not sure. Tom is allowing no stone to remain unturned, it seems. Karen shows them genotyping results identifying the double transgenic pups. No one seems to find it odd that she shows this. Perhaps because they are all presenting heaps of uninteresting details at these meetings. Perhaps because only Tom is paying attention. Karen next turns to the cell culture experiments. Like for Juan and Hiroshi, her results so far match those in the paper. Karen reports this neutrally. It might have been satisfying, in some perverse, self-justifying, way if nothing worked, if Chloe had made up all the results. But Karen never thought this was likely. Cutting corners seems like something that Chloe could do, always knowing better. But making up results all together? Not the same thing. Chloe is a scientist, after all.

Tom is nodding along as Karen talks. He now knows that most of the published results are trustworthy. This gives him some measure of relief. But like Karen, he also knows that the problematic experiment has yet to be replicated. So it is still a wait and see situation.

Karen finishes up and Yuqi launches into her presentation. Karen sits down and looks only at Yuqi thereafter, appearing to pay attention to the presentation. As Yuqi's words and numbers roll over her, Karen's thoughts go back to the results she just presented. If the mouse experiment was not done properly the first time, then hers will be the first real test of whether the inhibitor works in the animal. As it seems to work properly in cell culture, it would be logical for this last experiment to work as well, to be 'correct'. Everything else fits, so why shouldn't this? Paradoxically, they may well be on their way to show the data in the paper to be completely correct. Whether corners were cut or not. Then what do we do, she wonders. If all results are 'correct', the others will certainly want to know why the data were questioned in the first place. The whole thing will seem such a waste of their time, the "whistleblowing" unnecessary. But still. She has been down this track before. What matters is whether Chloe did the experiment or not. If she cheated. But they are examining results, not conduct. Karen cannot help but hope that the final result will be off. She is slightly ashamed of herself for thinking this. It would be bad for Tom. And for the lab? Not clear, really. But it would vindicate the whistleblowing. She shakes her head, too much speculation. Ungenerous thoughts or not, she has to keep the work squeaky clean, do every step of every experiment absolutely correctly. In just another month or so, it will be done. She is simply paying her dues to Tom, just like the others. She wishes she could find a way to actually believe this.

The meeting is finally done, but has taken almost two hours. Tom thanks everyone with a nod and a "Next Wednesday, same time". They leave quickly, not dawdling to chat as they might after a regular group meeting. Karen hurries to drop her things off at her desk and heads for the stairs, for the Küstner lab, to check on the microscope and chamber. After that, it is back to the tissue culture room to pick up the dishes. Once she has the microscope all set, she needs to find the best samples, and get going with the manipulations and the recordings. She mentally goes through every step, to make sure she does not forget anything. No time to waste on mistakes. Finding the best sample area is important. And the manipulations are still a bit tricky. Complete concentration is called for. So it will take a while. No rush jobs. She briefly considers whether she should call Bill and let him know when she will be home. No, she told him she was likely to be late, that he should just go ahead and eat. She knows this is not what he wanted to hear. But calling him won't help.

Karen lets herself into the apartment very quietly so as not to wake Bill. It is unlikely that he will be in bed before ensuring that she is home safely. But maybe he will have dozed off with a book or in front of the TV. She can only hope. It makes her feel less bad when she finds him that way. Alert and obviously waiting for her is worse. There is light coming from the living room, spilling into the kitchen. A lingering smell of something nice, garlic, slightly burnt cheese, tells her he has been cooking.

"Is that you, Karen?"

"Hi sweetie, yes, it's me. Sorry I'm so late. The busses are so irregular in the evening. You wouldn't believe...".

Bill has gotten up from where he was and is now standing in the doorway between living room and kitchen.

"You could have called me. I would have driven down to get you."

"I didn't want to trouble you. Just because I am working crazy late again." She tries to ward off reproach with a weak smile.

"I'll nuke some lasagna for you," He says, "It's good."

There was no question in this and Karen knows better than to insist that it is not necessary. It would only make matters worse. Besides, she is hungry. Bill removes the foil from a plate and puts it in the microwave. "There's some leftover red wine on the table, as well. In think there is a good glass still in there for you."

"Thanks, sweetie." Karen steps over and gives him a kiss on the cheek. He is looking intently at the plate in the microwave. Karen knows he has something to say. But she does not want to force it. He seems thoughtful, rather than angry.

After Karen has eaten half of her lasagna, "wonderful" she assures him, and has had a sip or two of her wine, Bill starts talking.

"Listen, Karen. I know the work is important for you. But you are overdoing it again. You are at the lab almost all the time now, early morning until late night, every day. It is too much. I'm worried you will run yourself to the ground again. Remember just after New Years, you were totally down in the dumps. Taking a breather helped."

"I know, sweetie. But this is different. As you know, Tom has us repeating experiments from Chloe's paper. That takes a lot of time, but it is just for one more month. Things will get back to normal again afterwards. And my cutting experiments are finally working." Karen looks up, imploring. "I told you, right? It's working beautifully. The results look really good. So I'm really excited about this again."

"I thought you got scooped on that, so the timing didn't matter anymore."

"It still matters to me." Some defensiveness creeps in. "I need to get my story out soon. Even if I don't get a trailblazing, top paper, now that I have the cutting results I should still get a good paper in a good journal. At least JCB, I think, if I publish soon. And if I do it soon, it will be clear that my work is not derivative. Don't you see? The timing does matter. So I have to get the experiments done quickly. I just have to, even though Tom wants me to help fix Chloe's mess as well. I have to do both. This really matters to me. You understand that, I know you do."

"Of course I do. I know you. Maybe what I have most trouble grasping is why you have to push so hard to repeat Chloe's work. Why does that have to be done in such a rush? The paper is out already. And does it have to be you? I mean, doesn't Tom see that you have to finish your own stuff?"

"I'm not the only one. He put four of us postdocs on the 'repeat team', and Lucy too. And I can hardly say no. I sort of started it, didn't I?"

"No, you did not start it. She did. She started it by misreporting her experiments. She did wrong, not you." Bill says, adamantly. "You did the brave thing. You spoke up. Tom should be damn happy he has conscientious people like you in his lab. Not just fast-tracking careerists who don't mind a bit of misconduct when it's convenient."

"I don't think he feels very happy about my bravery right now. He would be much happier if it all just went away. I should just have kept my mouth shut." Karen says, quietly.

"Well," Bill frowns "that horse left the barn. But why don't you tell him you can't do it right now? You need the time for your own work. He'll understand. And I'm sure he's aware of how bad it could look if he puts too much pressure on the whistleblower. That's a complete no-no."

"Bill, I'm just doing what the others are doing. I'm not under more pressure than they are. He is not singling me out and he has not told them that I was the one who brought this up. Imagine how they feel, being forced to push their stuff aside to do this. At least I know why. Anyway, I just want to get it done, finished. I also sort of understand Tom's argument for this. To get at the truth."

"Oh, come on, Karen. You know it's not about all that idealistic stuff. Or even if it is, it's not just that. Tom wants to save his own ass. And he is using you guys to do it. I don't understand why you feel so much loyalty. Why doesn't he just retract the damned paper and let the rest of you get on with what you are supposed to be doing?"

"You're being too harsh. We are not being used. We were asked and we agreed. Retracting a major paper hurts the whole lab, not just Tom. If Tom's reputation is tarnished, ours is too, by association. Anyway, so far it looks as though most of the paper is correct. Maybe all of it will be. I don't know." She adds the last drops of wine to her glass after Bill shakes his head at the raised bottle. She sighs. "I'm no longer sure I know what is right. But I am involved and I feel that I need to help sort it out."

"Just because you noticed and you let Tom know? That does not make it your mess to clean up."

"No, because... Well, I don't know. Tom was really trying to help me when the Science paper came out. I was down, and his support, his seeing that I have something to offer, really meant a lot to me."

Karen knows she is not being completely forthcoming. She cannot tell Bill about the other reason, about the sinking feeling she gets whenever she passes Chloe's bench, the shame, or guilt or whatever. She may never be able to explain that.

"Don't kid yourself. Guys like Tom think about themselves first, and everyone else after. That's how he got to where he is now." Bill says, with some real bitterness in his voice. "They don't care how many others they have to step on to get to the top—or to stay there."

"It's not like that." She counters, lamely. "He is a good guy, really."

"So you are not going to say anything to him? You are just going to keep on working 16-hour days?"

"Just for now. It'll get better in a month's time, when I finish Chloe's experiments. I promise." Karen is keenly aware that she has said something very much like this before, but she has nothing better to offer. "I am really tired. Can we go to bed? Please? As soon as I have finished this very yummy lasagna." She nods at the food left on her plate. "I really do appreciate it, you know. You worrying about me. And my dinner. I do. Really"

Bill looks at a darkened window. It seems he has said all that he is going to say for now. He probably does not want to push too hard, to be a nagging husband. He knows that Karen is tired and stressed. And who can resist his cooking being complemented with a pleading smile? For a while he stays where he is, gazing out, fidgeting with his empty wine glass. Karen finishes her lasagna and the last of her wine. Bill stands up and comes around the table, behind Karen's chair. He leans over and wraps her in a hug. She accepts, gratefully.

"Yes. Let's get some sleep," he finally says.

Chapter 13

Spring will soon be here. She can feel the promise of it. This morning was cold, with frost in the air. Now, at mid-day, it is still chilly, but the sun is strong whenever it shows itself. Turning a corner, she feels the warm sunlight on her wind-blown face. She also sees her apartment building up ahead, its stark concrete facade not yet softened by leaves in the trees out front. Only a couple of blocks to go now. Her legs are tired, but they still obey her. A satisfying awareness of the muscles in her thighs and calves has been with her the last couple of miles. Before that she had the magic miles of flying, soaring, feeling free and strong. She could go on forever.

Reaching her building, Chloe starts her stretches while still outside. Going inside immediately would be stifling. Too warm, not enough air, not enough space. Better to stay in the open, unconstrained air. Two winter-wrapped women with baby-strollers walk by, glancing at her as they pass. A different species, they are. She continues stretching until the sudden cooling of her skin tells her that the sun is in hiding again.

She climbs the stairs at a measured pace, noting, as last time, the minor differences at each floor. During the past month she has seen more of this building in daylight than all of the previous 4 years combined. It lacks the charm of sharing a house in Cambridge or Sommerville, like some of the other postdocs do. But it is adequate. And she can keep herself to herself. No busybody housemates. The spare neutrality is starting to get to her, though. It is not an interesting place.

Apartment 4-07 is a small one bedroom, with a kitchenette nestled off in the corner of the living room. Once inside, Chloe takes a bottle of cold water from the fridge, pours and drains, one glass after another. There are few pleasures as basic, as satisfying, as cold water after a long run. It is something to be appreciated. That, and the post-run purring of her body for the next hour or two. She sheds her running clothes and steps into the shower. The water is almost scalding to her still-cold fingers. After a moment, she relaxes and enjoys the water rolling over her shoulders, back, legs, arms, face. She keeps turning up the heat, bit by bit, as her body adjusts. Finally it is hot enough. She turns off the tap and reaches for a towel to wrap her red-splotched body before it cools again.

For the afternoon, her plan is to work on the review article and the grant. With lunch in hand and a cup of tea, she settles down at her desk. She should finish up the review first, she thinks. It has been in the works for a while now. She wants to get it completely ready for when Tom comes around and things return to normal. Most days, she is eager to get going on her first RO1 grant application, instead. It is for the future, so she is drawn to it. She has the basic ideas for the aims and the experiments pretty much worked out already. They will need to be carefully argued and described in detail. For the introduction part of the grant, she may be able to use some text from the review, so, two birds... Well, there is plenty to do. Working at home is quite productive, she has found. No distractions. It is not possible to take refuge in yet another experiment. And cleaning is a self-limiting displacement activity in apartment 4-07. Naturally, she will not be able to finalize the grant until she knows where she will be working next year. Most of the time, she manages to ignore the thin voice saying 'if' she will be working, 'if' she has a place next year. It is not helpful. She has to stay positive. It will all work out, it will all blow over. She will soon move on to her real life.

Today, though, she is having a harder time than usual suppressing the doubts about the future. She knows that having more preliminary data would make her grant stronger. But to generate the data, she needs to be back in the lab. When she was called to the lab yesterday, she thought that was the reason. It was all over. Tom was going to tell her everything was fine and she could get back to work. In the hours after Deidre called her to set up the appointment, she was cautiously optimistic, almost cheerful.

She is not naïve, of course. She knows it will be awkward to go back to the lab after all this. But she will manage. If anyone questions her about her absence or even about the investigation, she will be diplomatic. She will explain what happened objectively and straightforwardly and without assigning blame, even if she now has a pretty good idea who started all this. She is convinced that this magnanimous approach is the smartest one. There is no proof as to who it was, after all. And she just wants to move on. Michel, Juan, Vik and other friends in the lab will understand and be supportive, she thinks. She hopes. She will do her best to dispel any discomfort on their part. It would be tempting to hint, gently, at the malicious act that caused all this. But better not to. Lab friendships are temporary, superficial arrangements that should not be challenged with such sharp tests of loyalty. She feels the truth of this, intuitively. Interactions with other lab members, with those that have been all too ready to think the worst of her, can simply be minimized. Avoid them, pretend these small-minded people do not exist. They will most likely oblige by avoiding her as well, she realizes. That will be fine. It is, after all, only for a few months, only until she can move on and start her new life, her real life. She is so ready to move on.

But the visit to the lab yesterday did not turn out as she had hoped. The investigators were not finished. They needed her to show them—again—where certain sets of data were kept on her computer. And they wanted to see more raw data. They seemed to be obsessed by photos, data files. That was what the visit was for. There were no answers, just more questions. Actually pretty much the same questions that she had been asked earlier. It was also the same set of investigators, an older man and a younger woman, both formal and polite. They introduced themselves, again, but Chloe did not pay attention to the names. The lawyer-with-a-PhD was not with them this time. So, had the investigation now been delegated to underlings? These two did not inspire much confidence. Her impression was that they did not understand the science. Interacting with them, answering their simple, narrow questions, was frustrating. She wishes they would mount a direct challenge, so she could prove herself outright. Instead, it is this endless focus on minor details. She met them in the meeting room close to Tom's office. Her desk computer was there already. It had been set up and was turned on. The IT administrator access would probably allow them access to everything even without her help. But they asked her to log on, so she did, and she found the requested files for them. She noticed that they had her notebooks, as well. The numerous colored paper slips inserted in one of them suggested they had been through it in some detail already. Some of their questions were also more specific than previously. She explained, again, what she had done, which data she had kept, which not. And then, abruptly, it was over. They had no more questions.

She saw Tom afterwards, briefly. It was not very helpful. He could not tell her when it would be over. He did update her on the repeat effort, that bizarre lab endeavor to redo some of the experiments from her paper. If it takes 4 or 5 postdocs to do it, at least Tom should appreciate what an enormous effort it was for her. But the others must be annoyed at the imposition on their time. Chloe knows that she would be more than a bit irritated if she was asked to set her own project aside for something like this. This will undoubtedly add to their dislike of her, later. They should blame Tom for it, however, not her. He claims she suggested it, that she should welcome the effort to validate her paper. But thinking about this team in action infuriates her. These are her experiments, her ideas. But now the ideas, the project, the results all seem to belong to Tom. As if they would even exist without her. It is just plain wrong. Of course, Tom will only hold on to ownership if everything works the way it should. It will. She was not surprised to hear that the experiments so far all have the expected outcome. They get the same results she got. Quelle surprise!

Well, it does reassure her a bit. At least they are not screwing up the experiments by inattention, incompetence or even deliberately. This could happen all too easily. Tom has not told her who is on the repeat team but she knows this from Juan. They met at a tournament a couple of weeks back. He seemed a bit uncomfortable but she tried her best to be natural and pleasant and to get a conversation going. Finally, he loosened up. After some small talk, he started to tell her what was going on in the lab. When he understood that she knew about the repeat effort, he told her more about the team. It became clear why he was so uneasy about it. Not only is he on the team, but he thinks the whistleblower is, as well. He told her about the suggestion that Vik had made to Karen at a group meeting while Chloe was away, right after Karen had been scooped so badly. He suggested using the Myc and Ras double transgenic mice for some follow-up experiments. So it was not hard to work out who would have been talking to Andy about these particular mouse strains at that particular time. But that was not all. Juan had also seen Karen snooping around Chloe's desk one morning while Chloe was traveling. He had not told Tom about this, not knowing what to make of it. But he clearly needed to tell someone. He kept his eyes on the game being played in front of them while he revealed all this. He apologized, but it was unclear what for.

Chloe was convinced the moment Juan mentioned Karen's name. Karen, yes, of course, she must be the whistleblower. She seems to be the furtive, envious type. With Chloe's success and her own problems, it all made sense. That she had been sneaking around in the early morning, going through her desk, that was a bit shocking. But she tried not to alarm Juan. She did not want him to feel worse. Why Tom would put the whistleblower on the repeat team is a mystery. She could easily screw up an experiment, get a different result and then claim Chloe must have made a mistake, originally. This way she could justify herself for making the allegation in the first place. Such a scenario does not seem far-fetched. But pretty terrible. Tom cannot want a retraction, however. So why set it up like this? It made no sense. Anyway, as it was, Chloe could do nothing with her knowledge, least of all confront Tom with it. But the unasked questions and the suppressed background information added to her inner agitation in his office yesterday.

Tom's update on the experiments repeated thus far was just the straightforward facts, without questions or comments. She could only nod along, her "I told you so" implicit. She wasn't sure if there might not be a hint of apology in his voice. Or was she deluding herself? They both acted neutral and cordial, more restrained than in previous discussions. To an outsider it might have seemed like they were discussing something perfectly normal, the recent results of a lab project. But of course it was not a normal discussion. He seemed as uncomfortable as she was, as keenly aware of things that could not be said. Naturally, he did not mention the whistleblower. She did not repeat her accusation, which would now have a name attached. She felt almost ready to explode, but actually said very little. They blandly agreed to talk again when there was more news from the investigators or from the repeat work.

Just as she was getting up to leave, Chloe remembered to ask: "And the review, when is it due?"

"The review, what review?"

"The Annual reviews one. The one I started writing while I was traveling. They asked you for a contribution. And you asked me. Remember?"

"Oh. Yes. That. I'd forgotten all about that, I have to admit. But, don't you think? Maybe it is better to wait and see?"

"It is almost done. I'll get it to you soon."

"I'm not sure that would be . . ."

"I'll send it to you. Soon."

Tom looked uneasy but did not respond. Instead, he looked at his desk, at the computer screen. Chloe remembers taking a deep breath and making herself leave the office quickly, closing the door firmly behind her.

After the strained meeting with Tom, she felt a strong desire to leave the building as fast as possible. She could have gone back down the rear stairs and avoided the lab. Instead she more or less automatically turned down the usual corridor. As she passed the lab, she could not help but look through the glass windows, into the fish tank. Her eyes where drawn to Karen's bay. She was not there, not at the bench, not at her desk. Is she avoiding me? Chloe thought. Well, she would not know about the visit today, would she? Maybe she saw her go into Tom's office and is now hiding out of sight. A coward. Karen should have come to her directly, with her doubts and questions. She should have shown some respect. Running to the boss was childish or underhanded or both. And sneaking around Chloe's desk when no one was around, what was she up to with that? It makes her so angry. She is a devious sneak, a jealous bitch. Chloe reminded herself that she could not be sure Karen was the one. But her gut kept telling her that she was. She had to be.

Chloe moved on, but more slowly, still scanning the lab. Juan was at his desk and glanced up as she walked by. He made eye contact and gave a nod and a tentative smile. His expression seemed to be one of understanding. Luckily he stayed where he was. She did not feel up to discussing why she was there, even with a sympathetic listener. Behind Juan, her old desk looked incongruous. Her computer had been removed and there were gaping holes where her notebooks should be. The bench looked relatively undisturbed, but clearly empty of activity, dust settling on untouched racks and reagents. Plundered and abandoned, like foreign territory—strange to think that this bay was her whole world just a short while ago. This is so wrong, she thought. It is mine. I will get it back again. She quickened her step. A couple of other postdocs were in the lab but preoccupied and did not see her. This was probably for the best. Enough looking in. Keeping her distance to the lab for the time being was the smartest thing to do.

Oddly, she did not feel quite ready to return to the world outside. She decided to go upstairs instead. Without analyzing the thought too carefully, she convinced herself it would be nice to pass by Martin's lab. On the fifth floor, everything looked normal. Nothing much had changed since last time she was here, as far as she knew. There were no recent departures or new fellows. As usual, concentrated activity was evident in all of the small labs she passed by. Everyone seemed so sure of what they are doing and why, she thought. This is how science should be, full of determination and dedication and focus. She found herself smiling, dryly. They are so lucky, she thought. The fellows, they are their own masters. They have no lab mates or lab heads to control their fate. She wished again that she had gotten one of these positions instead of a regular postdoc. Martin had come later. There had been a vacancy then. He was lucky.

Martin's space is halfway down the corridor. As she got closer, she saw the back of his head in front of one of the desktop screens in his office, away from the corridor. Even as she reached the glass door to his space, she could not see much of his face, only the glow of whatever was on the screen reflecting off of it. He was sitting by himself and looked engrossed in whatever was in front of him. He tilted his head slowly as if trying to get just the right angle on a challenging set of data. She has seen him do this many times, when considering a puzzle of some sort. Instead of entering the lab, she stayed in the corridor, looking in. There was no reason to interrupt him, not really. For some time she just stood there, undisturbed, and gazed at him, equally undisturbed. She might have looked out of place. After all, no one usually stands still in the corridors here. So when one of the other fellows walked by, he offered a curt "May I help you?" to determine what she might be doing there. She recognized him, but did not remember his name. He apparently recognized her as well and mumbled a quick "Oh, Chloe, Hi, nice to see you" before he was off again, leaving no time for even a cursory response. The small disturbance was enough to catch Martin's attention, however. He glanced up and saw Chloe. The look on his face when he saw her was unmistakably one of displeasure. He corrected himself quite effectively, donned a smile and started to get up. But she had seen it. She flashed a smile and signaled for him to stay where he was. Adding a wave, she was off, in a hurry, back down the corridor.

Thinking about yesterday, she wishes she had not gone to the fifth floor. She cannot imagine why she thought it was a good idea at the time. It is all so complicated now. She has not talked to Martin since last weekend, but she expects to see him tonight. She is not sure whether she even wants him to come over. Maybe she should cancel. The whole visit to the institute yesterday was disconcerting. She wishes she would never have to go back there. But that of course makes no sense. She will need to go back to the lab, just for a while, to continue her work. She should get that last bit of preliminary data for the grant. And she most definitely does not want to be driven off. Even if the only thing she really wants is to start fresh, far away, as soon as possible. With that thought on her mind again, she checks her email inbox yet another time. Every day she hopes that it will contain a job offer, or at least an invitation for a second visit. It is still relatively early, she knows, and most places may not have finished the first round of interviews yet. But she wishes she would hear something. She needs to grab on to the future, to pull herself away from this old place. She needs a future to grab on to.

The inbox contains nothing that looks like an official correspondence, but one new mail that catches her eye. "SF here you come???" Sent by Barry less than half an hour ago, so still morning on the west coast. He might still be at home, as he seems to be a late worker. Maybe she should call him. She and Barry have been exchanging emails and phone calls since shortly after she got back from San Francisco. It started out as a friendly and straightforward chat. It seemed a natural consequence of their good rapport during her visit. Barry has been doing his valiant best advertising San Francisco as the best place for her to start her own lab. He adds a new reason every week, sometimes serious, sometimes funny. The science scene there certainly seems to be lively, with some of the most creative minds and lots of energy. Then there is the city itself, with its life and charming hilly streets as well as the wine country. And people are nicer, he claims, not as rigid and pompous as on the East Coast. Somehow it all adds up to creating a fantastic place that no one ever wants to leave. She has never lived there and does not really understand the hype, but he is so enthusiastic that she lets herself be carried along.

Most importantly, Barry is confident that she will get the job. And his confidence is infectious. He has been to all the job talks and many of the dinners. He remains certain that she is the frontrunner. Apparently, the last interview is this week, so they may be making a move soon. Along the way, he has told her who the other candidates are and any extra information he has been able to dig up. Although he is not on the search committee, he seems to know a lot. He has even found out which candidates some of the senior faculty may have an interest in for less obvious reasons. There is no 'insider candidate' as such, but someone could be an optimal collaboration partner for existing faculty members or fill a perceived gap in the department's knowledge. He shares all of this scattered information with her and they pour over it carefully. From the names and affiliations, she can put together the CVs of the other candidates, more or less, and evaluate them, on paper. Based on this, she allows herself to share his optimism. She does look good, indeed. And it is fun, in a slightly guilty kind of way, to dissect and discuss the competition. Alongside this analysis of contenders for 'her' position, they have been exchanging additional bits of personal history and getting to know one another better. Flirting, really. Flirting quite a bit.

The most recent email from Barry is short. He gives some information about this week's candidate, nothing secret or compromising, just who it is, the title of the talk and so on. They can get into the details when they talk on the phone. She should get this new thing, Skype, so they can talk for free and with video. But for now, she likes the voice-only communication. A voice can carry so much more when it is all you have to focus on. And she does not feel like worrying about the phone bill right now. She calls his home number and he picks up immediately.

"Hello"

"What are you doing still at home, lazy bones? It must be past 9 already"

"Burning the midnight oil as usual last night. I'm working hard here. What about you? Are you home as well?"

"Well, you know, working on my grant, and on a review. It is much easier to work here than to do it at the lab. There is too much disturbance there, too many people yapping away."

"Soon you will have your own office so you can close the door."

"Heavens. It will be good with a bit of privacy. For work, of course. And for gentlemen callers, as needed."

"Gentlemen? I thought I was the only one."

"Yes, but I might expand. You never know"

"Hmmm. We can discuss that when you get here. I might have a few objections to that. Oh, and I walked by our bar on the way home. I almost stopped in. In just a few months we can drop by whenever."

"Assuming I get the job."

"Of course you will get it. You are the shining star, Dr. Varga. Your performance was by far the best. They would be crazy not to hire you. Plus I heard what they said after your visit. Remember? I told you they were over the moon. Really."

"So, tomorrow is the last candidate?"

"Yes, and she is not very serious competition. Her CV is nowhere near as strong as yours. I more or less worked it out already. Her name is not a very common one, so I just looked her up on Pubmed. It looks like just one person with that name, one set of publications. One rather limited set, I should say."

"You went through the publications already? That's usually my job."

"What can I say? I was waiting for your call and I had some time. Plus I am getting impatient. So let's say the committee meets right after her visit—or at the latest next week. You should have your invitation for a second visit soon."

"Assuming all goes well."

"It will. Don't worry. Remember, I have seen all the candidates. I know everything worth knowing. You were by far the best. It's that simple."

"Simple?"

"Yep. So don't fret. Start worrying about getting yourself out here. The sooner, the better. I have all kinds of plans for your next visit."

"And those would be completely decent plans, I presume?" She says, switching back to the more playful tone. There is a short pause. She can feel him smiling.

"Not at all. I have a few indecent plans in there as well."

"Hmm. You are a bad influence, you are. I am not sure I can discuss such things right now. I might get a little overheated here. It is a pity that you are all the way out there."

"Maybe I can help you out, anyway. I do my morning emails in bed. And morning phone calls."

"You are so bad."

"Me?" He replies, with mock surprise. "Are you still having trouble with the temperature there?"

"Absolutely."

"Let me suggest something. . ."

After the call, she remains in bed for a few moments, warm, relaxed. She gets up slowly and goes to make some more tea. She had better focus and get some work done before the whole day is shot. Martin is coming tonight, she remembers. Well, it's OK. She will be in a good mood still. They can order some take-out, maybe a pizza.

Martin arrives with the pizza as planned. They are both hungry and dig in, immediately. Usually, the talk flows just as quickly. But not tonight. Martin seems preoccupied and withdrawn.

"So what's up? You seem so far away." She finally prods him.

"Well, it wasn't the best of days, to be honest". He reminds her that one of his closest competitors visited the institute to give a talk today.

"Oh, yes. I saw that on the website. Maybe I should have come in for that. As far as I remember, she is a bit of a tough one, but does interesting work, right?"

"Well, I suppose it was good that you weren't there, given how it turned out. You missed me getting stupidly agitated and pissed off. I am sure it was clear to everyone that something was up, but maybe not what. For once I did not ask any questions. They must have noticed that I was uncharacteristically quiet, despite the fact that subject was so obviously close to my work."

"But why?"

"Oh, you know. It was the usual devious approach to undermining the competition. She avoided referring to my work whenever possible and deliberately obscured where some of the key findings—mine of course—had come from. She was never directly untruthful, but just skillfully misguided the audience, very much the smooth operator. It made me so angry. But there was nothing I could say that wouldn't just have made me look like a nit-picking jerk. So I said nothing. But I'm really angry about it, still."

"That's really not OK."

"I know. It wouldn't surprise me if she does this even more blatantly elsewhere, but today she was on my patch. I couldn't help but feel that she was deliberately trying to make me look insignificant to my colleagues, to discredit or at least devalue my work. Hopefully they didn't pay too much attention. Or they will forget it again."

"Come on, they know you here. They know how original you are. That is why you were hired. Trying something like this just makes her look ungenerous. Or worse. If they think a step further, they'll realize she must be afraid of you, or why bother."

"I hope they see it that way. But the truth is, I could use a Nature paper right now."

Chloe does not respond, but saying this out loud appears to remind Martin who he is speaking to. He looks at her, no longer tense and upset, but questioning.

"So how come you didn't come in, yesterday? I saw you in the hallway."

"You looked very busy. And I was just there for a short visit."

"So did you get everything sorted with Tom?"

"It's pretty much sorted. Not completely, though. So it is easier if I am not there, unless I really need to be." She smiles, reassuringly. "I have my review to finish and my grant to work on. This is my first review with a direct cancer focus, so there's a lot of extra literature to read. And it takes more time for me to get a good first draft. It would be hard to get any concentrated writing done in the lab. There's too much distraction. Remember that I'm just a regular postdoc with a desk. I don't have an office, like you do."

"Working at home makes sense, then, I guess." He says slowly, with a slight shrug. Then he lights up, suddenly. "I almost forgot. I got this really neat result yesterday that I wanted to tell you about."

Martin explains how one of his most risky projects may finally be paying off. It would be his first major breakthrough as a fellow. He is excited and happy again, the visiting speaker already forgotten. He gives Chloe all the details: ideas, experiments, results, interpretations. This has always been part of their pattern, discussing projects, in particular his projects. She loves this intellectual sparring and jumps right in, providing questions and suggestions. They have a good conversation, everything else pushed aside for now.

But, sensibly, they call it an early night.

Chapter 14

Karen is on her usual early bus, in her usual seat, lost in thought. The early morning sun is lighting up a pale blue sky. Here and there, the still gentle light makes its way down side streets and through gaps in rows of houses. Soon it will catch the white and pink magnolias and delicate green of the new leaves on city trees. A beautiful spring day seems to be on offer. She usually pays close attention to the little events of spring. It is her favorite season, even if it is shorter here than what she is used to. This year, there is little time to enjoy it, however. Today is destined for the same fate. But she is smiling. Perhaps the jumping rays of the morning sun are not being completely ignored.

But Karen's thoughts are already in the lab. Her private smile comes from her unmitigated satisfaction with a long-awaited result—and from the fact that she, for once, is looking forward to presenting her findings to the repeat team this afternoon. There was something seriously wrong with that experiment, after all. A couple of weeks ago, with the first mice, she got a hint of what was to come. Now, the evidence has accumulated beyond any reasonable doubt. The inhibitor has no effect in mice. It does not affect tumor burden as claimed in the paper. Not only were the numbers impossible, the result was simply wrong. It is completely clear. And she is happy.

She was eager to get going this morning. Bill did not comment on her rapid departure or ask any questions about her chirpy mood. He seems to have accepted the importance of this strange task to her, even if he doesn't like it. And of course he knows how important the progress on her own project is to her. They have sort of a truce. She works as much as she needs to for now, but she does not talk about it at home. He does not push for answers or change and tries to be content with the little time they have together. He sees old friends more often, on his own.

There is still a lot to do to get everything ready for this afternoon. Slides to look at and numbers to crunch. Having done the gross analysis and taken the required pictures with Andy on Monday, she is confidant of the basic result. The tumors were there in both sets of mice—with and without the inhibitor. But if she can wrap up the remaining analysis of histology data this morning, she can put a complete end to the repeat-team nonsense. Today. And, in the end, Tom will know she was right to speak up. It is a big relief. As all the other experiments from the paper got confirmed, it seemed more and more likely that this one might as well. Even if it had been done incorrectly the first time around. It would then be much harder to see her pointing out the problem with Figure 7F as a necessary, brave step. It would seem like a time-waster or even start to look suspect. Chloe would be vindicated, in a way. Both aspects would be too embarrassing to bear. So getting this negative result was very welcome. It made her happy and it still does. She admits this to herself, knowing that it is ungenerous. But she has not told anyone yet. She is not quite sure why not. She is looking forward to revealing it this afternoon, that is true. Tom may get upset, but he will just have to accept it and do what he needs to do to set this right. Tom's problem, not hers. From now on, she will focus all her energy on wrapping up her own paper. The other repeat-team members have already returned to their own projects, having completed their assigned tasks. The team meetings have therefore become shorter but also, for her, tenser. Today will be the last one, she hopes.

Walking from the bus stop to the institute, she remembers that today is also the first day in a long time that she will not go to the second floor microscope as the very first thing. She did not set up a recording yesterday. Monday evening she did, but she was so distracted by the inhibitor results that she did a poor job and it was useless. She has been continuing with the cutting experiments beyond what was necessary, anyway. She had enough data to make a solid case a week or two ago. But in some strange way it seemed important to continue the work in parallel, not to swerve. She kept everything going until both her penance on the repeat team and her formerly intractable experiment were done. So that one wouldn't jinx the other, perhaps. Superstitious and completely silly, she knows. But that doesn't matter. It is done now and she can move on.

As always, for her early morning entrance, one of the security guys buzzes her into the building. Even the new ones get to know her by sight pretty quickly, the always-early bird. She hurries directly to the third floor, to the empty lab and the samples awaiting her attention.

"So what you are saying is that there is absolutely no difference in tumor number or size between the mice that received the inhibitor and those that did not?" Tom repeats, not trying to conceal his skepticism.

The numbers on the graph are perfectly clear, as are the accompanying grisly images of mice splayed open to reveal numerous tumors along the body wall. The three on the left are MMTV-Ras plus MMTV-Myc mice treated with inhibitor, the three on the right are the matching controls. They all look the same. As they did on the last slide.

The repeat team is in the small meeting room as usual. It is only four twenty, but they are almost done as only Karen is presenting. Although she had been looking forward to this moment, the slight unease she felt an hour ago seems to have been warranted. Only now does she appreciate that she is the only one with a reason to be happy about the result. She had not fully considered what it would mean for Tom. He said he wanted the truth, but maybe she should not be so surprised that he finds it hard to accept the result. Maybe she should have told him in private first, to give him time to get used to it.

"Yes. That's what I am saying. To be on the safe side, it was all done blindly. I scored number of tumors per mouse by gross morphology without knowing which mouse was which. Andy had coded them for me. Not that it actually matters, in hindsight, since there was no difference, as you see." She points to pictures of the mice again. "You cannot tell them apart. I looked at the last mice Monday, and decoded tumor scores then. The histology samples were also coded and analyzed blind. I finished the analysis today. Same result. The tumor area and grade are very similar with and without inhibitor." She changes to the next slide and points. "If there is any difference, it is a slight tendency for the tumors from mice on the inhibitor to have a higher grade. So they are slightly worse off, not better. But the difference is not statistically significant, so I wouldn't take it seriously at this stage." Throughout this, Karen maintains a neutral tone, treating Tom's remark as a normal question rather than a challenge.

"Are all these numbers from one experiment, done on one day?"

"No. They are from six mice per condition, half maternal Myc, half paternal, sacrificed on three different days. Matching sets of experimental and control mice were done in parallel, according to age and drug schedule. Only the decoding was done all together, at the end."

"That shows a surprising amount of restraint. Weren't you curious about the results earlier? Couldn't you have told us the preliminary results a week or two ago?"

"Of course I was curious, but I thought you told us to blind the experiments wherever possible. Especially for experiments like this one, that take a long time to do and are not easy to set up again. And I wanted to be sure."

"You did the right thing, Karen, of course. But. . ."

"I did suspect this would be the answer already a couple of weeks ago, given how similar the mice looked. I just didn't want to jump to conclusions."

"Yes, of course, of course. It is just a bit sudden, this negative result. And, well, it is also the first and only negative result we have seen. I am just saying, it is surprising given all the other experiments have been OK. Even the tissue culture experiments with the inhibitor, they worked. We just need to make sure that there is no mistake here."

"I know all that." Karen answers, coloring visibly. "I did most of the tissue culture experiments with the inhibitor. I know they worked. But like it or not, this is the result Andy and I got with the mice. You can ask Andy if you don't believe me. We had six mice in each group. There is no difference. The result could not be clearer. This was also . . ." She stops herself, remembering that the rest of the team does not know which result from the paper was questioned in the first place.

"I'm not attacking you, Karen. I just need to be absolutely sure what the correct result is. You are saying that the published result is wrong. Did you use the same dose as in the paper?"

"The same dose and the same timing, yes. And the batch of inhibitor is the one that I used for the tissue culture experiments. It worked there, so the inhibitor is OK."

"Right, I am just saying, you know, negative results can be difficult. You need to be extra sure every step of the experiment was done correctly."

"I know that. The experiment was done exactly as we discussed, exactly as described in Chloe's paper. The only difference is that I had a larger sample size, more mice. And as far as I know, that's a good thing." The last remark is gratuitous, she knows, but she cannot help herself.

"Well, yes, of course it is. I am just trying to make sense of it all. Why would this one experiment with the inhibitor not work."

"There could be lots of reasons." Karen says, heatedly, "Drug metabolism in the mouse, for example. Drugging an animal is not like adding a chemical to a tissue culture dish."

"OK, OK." Tom stops Karen with a raised hand. He then turns and faces Lucy. She is sitting off to the side and has, as usual, been quiet during the meeting. As Tom's long-term technician, she has been helping them out as needed, but has not presented results of her own.

"Lucy, what did you find? Do you have your results?"

Lucy looks down and does not answer immediately. Although older than the postdocs by 10 or 15 years, she lacks their confidence. For a moment, Karen is also quiet. She is confused. Why is Tom asking Lucy about this? How would she know about the results? A split second later, she understands. She turns to face Tom. When she speaks, her voice is controlled but full of indignation, slowly turning to real anger.

"Did you have Lucy do the same experiment? Did you really have duplicate experiments set up that the rest of us knew nothing about? How could you do that?"

Tom does not respond to Karen. He is still looking at Lucy, awaiting her answer.

"You ask us to spend all this time redoing experiments from a paper that none of us really want anything to do with—to help you—and then you don't trust us? This is outrageous. It is insulting." Karen is usually able to control her emotions in front of an audience. But she is failing completely here. Lucy seems unable or unwilling to speak up now. Juan and Hiroshi are both staring hard at the table, staying out of this fight. Finally Tom turns back to Karen and says, very evenly:

"Yes. I asked Lucy to do another set of mouse experiments. This experiment was particularly critical, for a number of reasons. And Lucy has prior experience with these mice."

"Was it just me, Tom?" Karen continues. "What did you think I would do? Or did you do this to everyone? Do you not trust us?"

"Look, Karen." Tom says. "Think about it. This should not be so difficult to understand. This whole business started with not knowing whether Chloe's experiments were done correctly and could be trusted. So whether you like it or not, I have to be absolutely sure that I get the right answer now. Simple trust is now a luxury I cannot afford."

"But . . ." Karen starts, then falters. It sounds so rational, put this way, but it still feels so wrong. Tom did not say whether it was just her experiment being shadowed, or everyone's. She is not sure she wants to find out. She has already drawn too much attention to herself.

"In any case, you have nothing to worry about. I'm sure you followed the protocol and presented everything accurately. If so, then Lucy will get the same answer. Now lets move on." He finishes, tersely, and turns back to Lucy. "What did you find, Lucy? Do you have your results yet?"

"No, not the final results" Lucy falters, clears her throat. "My treatment of MMTV-Ras and Myc double transgenic mice got started later, when the next round of mice were available. So I won't get the final results until two weeks from now."

"So we have to wait and see, then", Tom says. Silence. The tension in the room is palpable. The thought of this remaining unresolved for another 2 weeks seems unbearable to Karen.

"Well, maybe not," says Lucy. "You asked me to test the inhibitor with other tumor models that we had going. We had a batch of newborn MMTV-Neu mice that I could start treating with the inhibitor. Along with corresponding controls, of course. Those, I have analyzed already. It is different from the published experiment, of course, but. . ."

"But what? What did you see?"

"Well, no difference. Like Karen showed us, the tumor incidence and size are the same with and without the inhibitor." She looks over at Karen, still standing by her projected slide. Karen closes her eyes briefly. She feels a surge of relief. Thank you, Lucy.

"Seeing that, I started to worry." Lucy continues. "So I did a preliminary analysis of the MMTV-Ras and Myc mice. I haven't dissected them yet as they have almost 2 weeks left to go according to the protocol. But I inspected them visually and I palpated carefully." Lucy has remained seated and is speaking softly. But all eyes are on her. Tom is visibly agitated. Lucy seems to be making an effort not to look at anyone. "I can feel tumors both with and without the inhibitor. I couldn't count or measure the tumors accurately, but I am reasonably sure that the two groups of mice are similar."

The room is completely quiet. Finally Tom asks. "So you agree that there is no detectable effect of the Jmjd10 inhibitor on tumors in mice?"

"Yes, as far as I can tell, there is no effect." Lucy answers, quietly. "I was going to tell you, but. . ."

"You wanted to be sure. I understand. There is nothing wrong in that." Tom pauses "So the inhibitor affects tumor cells in culture, but not in mice." He does not need to say anything further. They all understand. A key experiment in the paper is wrong. They have their answer. And Tom will have to deal with it. He starts collecting his things absentmindedly and getting up from his chair. In a far-off voice, directed at no one in particular, he says. "We are done here. Whatever remaining results you have, hand them to Lucy. All the data, please, and copies of your presentations. I'll let you all know what

I decide to do." He looks around the room. Everyone looks beat, subdued. No one says anything as they get up. Tom draws a deep breath and continues, now in a more formal tone of voice. "Thank you, all of you, for all the extra effort these last couple of months. I know it must have been frustrating, to have to put your own projects second. But I really appreciate your contributions." They nod and mumble their replies as they leave the room, one by one. Only Lucy stays behind, still seated.

"I am so sorry," she says quietly. "I understand what this must mean for you. I should have told you earlier."

"Don't be sorry. You did what you were asked to do. Everyone here did. I should probably be the one apologizing, for putting you in this position, for making you keep your experiments hidden from the rest of the team. But that was just the way it had to be. I couldn't take the risk. Now I just have to deal with the fallout. You have nothing to worry about. Really, nothing." He manages a smile. Lucy nods and leaves the room.

Karen is waiting just outside the seminar room. Assuming she wants to talk to Tom, Lucy moves on. But Karen stops her.

"Wait, Lucy. Look, I'm sorry I got so upset in there. I just wanted to make sure you know that I am not upset with you. You did what Tom asked you to do. I understand. I know it was not your decision. I just think it was shitty of Tom to do it this way. I am shocked that he didn't trust me, and I'm angry at him. But not at you. So are we OK?"

"Of course." Lucy's pale eyes seek out Karen's. She smiles tentatively and, seeing a smile in return, relaxes. As she continues talking, Lucy's voice gradually regains its normal pleasant eagerness. "I'm sorry, too. I felt really bad having to keep it secret. But I figured that Tom must have had a good reason. In any case, it all turned out OK, didn't it? I mean, it looks like we are getting the same result. Even if it is not the result Tom wanted, at least it's consistent. Right?"

"Yes, that's a relief, I admit. Being the only one with a negative result would have been awkward if Tom wasn't prepared to believe it. And it's good that you have the MMTV-Neu data as well. We can be pretty sure now that the inhibitor does not affect mammary tumors. It's not just something unusual about the ones with Ras and Myc."

"But what should I do with my MMTV-Ras and Myc mice? If it is all over, as Tom said, should I sacrifice them now? Or when they were due to be analyzed anyway?"

"Maybe you'd better ask Tom. I think it would be safest to keep them going on the treatment for the time being. That way, if Tom wants you to do anything else with them, you won't need to start all over again."

"Right. I'll just keep them going until he says otherwise, at least until the tumor burden gets so bad that I'll have to euthanize them."

"I have more double transgenic mice coming along as well, if we need them. You never know, with Tom." Karen glances at the still-closed door to the seminar room and shares a long-suffering look with Lucy.

"So, Lucy, I should give you my data. I have everything on my thumb-drive here. Should we transfer it to your computer now?"

Having completed this job and having now also ironed things out with Lucy, Karen feels light, unburdened. For the first time in a long while, she feels no urge to get away from everyone as quickly as possible. Lucy is a nice person. She can be nice in return. There is lots of time to get on with her work later.

Once Tom is alone in the seminar room, he sits back down and rests his chin on folded hands. He stares at the heavily annotated paper in front of him, his "updating" of Chloe's paper. For every experiment related to Jmjd10 and the inhibitor, he has written the repeat team's result next to the corresponding published result. This has produced a dense array of neatly printed numbers and checkmarks in the margin of the paper. Always losing pens, he has used several during these proceedings. The changing colors of his scribbles mark the weeks going by. It is an odd document, but it is a history of sorts. He has never had reason to repeat a large set of experiments from his lab before. Seeing it line up in this way, results very similar but not identical, textbook examples of experimental and biological variation, has been gratifying. Still, not enough to make him forget the problems that started it all.

And now this. Damn Chloe! He was starting to allow himself to think it would all be OK, that the paper would be sound. Whatever Chloe might or might not have been up to, he would be able to fix it with a correction, and not need a retraction. Now he has to wonder how far she went with the deception. Even if she only did the experiment with two mice, she should have seen no difference. She would have known the inhibitor did not work. Did she willfully ignore the evidence? Or maybe she did not do the experiment, at all? It does not matter, really. But damn her!

He knows that part of the reason he is so upset is that he was taken by surprise. Disappointed, of course, that too. But he would have expected Karen to tell him as soon as she had a hint that the experiment was going in this direction. She could, like Lucy, have examined the mice earlier, to get a preliminary result. Well, maybe not. He remembers that Karen blinded the experiment, so she wouldn't know which mice to look at. But what restraint! It is difficult to understand. He doubts he could have waited until the very end to check the mice, himself. At least a superficial examination. And he did not expect her to spring it on him like this. A visit to his office as soon as she knew, smug, vindicated, that would make more sense. She could have told him

Monday. Why did she wait? She must have felt that she needed to present this unwelcome data in public. Did she think that he would try to suppress the data? No, he could not do that. She would still have the data, no matter how he reacted. And surely she knows that he wants the truth. Has he not made that clear? Maybe she was afraid that he would get angry and decided to do it in the group for safety in numbers. That seems plausible. As it happens, she got angrier than he did, when she realized that he had Lucy doing parallel experiments. But he had to do that. She was the whistleblower. He could not possibly have let her be the only one to do the key experiment, the experiment that she directly challenged. Maybe he should not have had her on the repeat team at all. But with her expertise, it would have looked strange if she had been excluded. It would practically amount to naming her as the whistleblower. Anyway, what is done, is done. At least he now knows that she has been honest. There were also Chloe's counter-accusations to consider, even if he never actually believed her story of the stolen slides. She did claim that the whistleblower was someone who was out to get her. So he needed to be sure. He does not regret asking Lucy to duplicate the experiment. It was prudent. Who knows what those girls could come up with, when backed into a corner? Anyway, Karen will get over it.

Concentrate, he tells himself, concentrate. He needs to decide how to deal with the fact that the last result in the paper is wrong. Worse, it is wrong because of what appears to be scientific misconduct. It will be impossible to see it as an honest mistake. Everything else in the paper is perfectly fine. But that won't save him. He has not said anything to the journal, not yet. Spending an extra couple of months checking everything seemed sensible, initially. He could justify the time required for it. He still can, at least to himself. But could it give the impression he has been trying to avoid facing up to facts? Or trying to cover up Chloe's misconduct with new data? He is not, of course, but any suggestion of this could hurt him. He was not party to her misconduct. He did not do anything wrong. This needs to be made clear before word of all this gets around in some other way. If he steps forward quickly, says what has happened, and retracts the paper, the mud will not stick for long. He will lose a great paper. People will notice, especially with a paper in a top journal. They will shake their heads, maybe even commiserate. They will, most of all, be relieved it did not happen to them. What if one of their postdocs fabricated results and contaminated their papers? This is everyone's nightmare. Knowing it could happen to them, they may not be too eager to throw stones. An apology—and then it will be over. His reputation will take a hit but not too bad. They know him. He has done plenty of excellent work in the past. There is lots of other good work coming along. He will get beyond this. As the corresponding author, he can simply retract the paper. He wonders again if he should consider a correction. Maybe the journal would let him withdraw just the mouse experiment, and not make a full retraction. Well, no, that is not how it works. And it is better to face the music now. Better not to risk backlash for trying to salvage something from this God-awful mess.

Right, he thinks, it is obvious what he needs to do. It is time to write the retraction request and get it sent off, as soon as possible. He looks up and is surprised to find that he is still in the small meeting room. He strides down the hallway to his office. Deidre looks up and seems to be about to say something but he beats her to it. "Make sure I am not disturbed the rest of the afternoon" he says, "I have to attend to something urgently. If anyone asks, tell them I can see them tomorrow." He feels he must take action immediately, to get everything set straight.

A couple of hours later, Tom is almost ready to send it off. He is more than ready to have it over and done with. And it will be, as soon as this last phone call comes in, from Sushma Nayar. She said she would get back to him quickly, this evening. Before calling Sushma, he called their chemistry collaborator Kumar Singh and explained the situation. He apologized profusely and explained why he had no choice but to retract the paper. They have known each other for years and collaborated on numerous projects. Kumar was understanding and sympathetic. If he was curious about what was happening with Chloe or her motivations, he kept it to himself. It was not a very long conversation and they said their goodbyes on perfectly good terms. It is never pleasant to be associated with a very visible retraction. But it is obvious from the phrasing that Kumar is free of blame. Tom emailed him the relevant part of the text from his draft of the retraction request. It sits on the screen now.

An experiment performed by C.V. cannot be reproduced. The original data to support the findings reported in Figure 7F cannot be located and new, independent experiments testing the effects of the Jmjd10 inhibitor L-334 on tumor growth in MMTV-Ras/+, MMTV-Myc/+ mice did not reproduce the findings reported in Figure 7F. All other experiments involving Jmjd10 or L-334, as reported in Figures 3 through 7, have been repeated and successfully reproduced by other scientists in our laboratory. Given the significance of the result in Figure 7F, however, and in the interest of the public good, the corresponding author (T.G.P.) hereby requests that the paper be retracted in its entirety. T.G.P apologizes sincerely for any problems or inconveniences this may have caused.

From the editorial office they will want to hear more about the events leading him to send this email. But he is sure they will agree to his request to retract the paper and they will probably act quickly. Kumar will send a separate email to indicate his agreement. That the actions are being taken without direct input from the first author will be understandable, given the circumstances. Tom did consider calling Chloe before contacting the journal. He could also talk to her about it, face to face. This would perhaps be the more correct thing to do. But he has absolutely no desire to talk to her right now. He does not want to hear her excuses. He is not sure he trusts himself not to yell at her or say something he will regret. For now, he has had enough. He just wants to get on with it. To be sure of his legal position, though, he called Sushma to get her input. He also needed to check with her whether a retraction would affect their investigation of Chloe.

In their conversation earlier, Sushma was pleasant but a bit reserved. She did not comment when he told her of the lab's efforts to repeat all the data in the paper. Fine. It is his prerogative to make those decisions. She did go through the wording of the retraction notice with him and suggested some minor changes. With no direct mention of misconduct, she found it acceptable. She agreed with him that the paper should be regarded as a piece of work from the lab for which he is ultimately responsible and therefore could be separated from the misconduct case. But just to be sure about all the possible angles, she wanted to run it by her colleagues.

"I'll be quick, I promise. But if you could hold off sending the email until I get back to you, I'd appreciate it. I'll call you later today." She did ask him whether he had discussed the retraction with Chloe but made no further comment when he said no, not yet.

"Tom Palmer speaking" He picks up after the first ring.

"Hi. Sushma Nayar here."

"Hi. Thanks for getting back to me so quickly."

"You're welcome. And thank you for checking with me before doing anything. I do understand how difficult this must be for you. I have asked around and we are in agreement here at the office. We have no objections to your retraction notice. So you can go ahead, if you think this is the right thing to do for scientific reasons."

"Thanks, I will. Although I hate to do it, I know I have to. It's important to let the scientific community know that something we published is incorrect. I have an obligation to do that. I just wanted to make sure that I wouldn't get tripped up in some legal matter." A pause, "And how about the investigation, how is that going?"

"I am sorry, Tom, but I cannot discuss the investigation with you, ongoing as it is. Procedures, you understand."

"Of course. I should have realized that."

"We will let Chloe Varga and yourself know as soon as the investigation is concluded. It'll just be another week or two, I believe."

After they ring off, Tom turns to his desktop screen. He pastes the final text into the email to the editorial office at Nature. He reads it over one more time. Send. It is done.

Chapter 15

"Dr. Varga, Dr. Chloe Varga?"

"Yes .." she says, drawing it out. She is not sure she wants to be Dr. Varga at the moment. She has been expecting a call from Martin, or from Tom, so she picked up. But so far, there is no word from either of them.

"My name is Christopher Turrell. I am sorry to bother you at home, but Dr. Palmer's secretary said that's where I might reach you."

"What is this about?" She can guess what this is about, all too easily. She saw the retraction this morning, less than a week after the email from Tom. He wrecks her career, her whole life and he doesn't even have the decency to talk to her. He sends a short email. And he sends it after he has contacted the journal, after he has made the decision. She was shocked when she saw the email, in all its cold detachment and superficial politeness. "Dear Chloe, .." As if he didn't know, perfectly well, what this would do to her. Well, first she was shocked, then she was furious. But she did nothing. What could she do? Nothing. Over the weekend, she talked to Barry. They had a nice chat, as usual. Somehow she could not bring herself to mention the email. She saw Martin as well, and did not mention it to him either. Maybe if she just ignored it, it would go away. No such luck, of course. The actual retraction was published online much quicker than she expected. She saw it on the Nature website this morning. She saw it and just stared at it, completely numb. She has not done much since, not been outside, not talked to anyone.

She decides not to wait for the answer, "Is it about the retraction?"

"Yes, yes it is. I am calling to hear your side of the story, so to speak. If you are willing to talk about it, that is. I work with an online magazine called 'At the bench'. We do articles about science and the people behind it. Serious articles supplemented with online commentary, blogs for thinking people. We cover different areas of science, but the majority of articles deal with the biomedical side. I come from the this area myself."

"And why do you want to hear from me? What is this really about?" She won't hesitate to hang up, should she feel like it. Politeness is not a major concern for her right now. But she is also a bit curious.

"Well, I saw the retraction in Nature, of course. Then I read the original article. Fantastic paper. It really is. And I know enough about lab work to understand that it represents a huge amount of work. So I look at the retraction notice again and as far as I can see, it's all about one small part of one figure, just one tiny part of that great piece of work. So I wonder, why retract a whole paper because of one error? It's also obvious that the retraction was requested by Dr. Palmer and Dr. Singh. But not by you, the first author."

"No. They did not ask for my opinion."

"Well, so, the way the notice is written, the blame is quite clearly being put on you. When I saw that, my immediate thought was that this was very one-sided and unfair. A person is being railroaded here. And that person deserves to get their side of the story out. I mean, the scientific community is not just the professors. It is also all the postdocs and the PhD students, who do all the work but get minimal credit. And then something like this happens. Something goes wrong somewhere and presto, the first author gets all the blame. You are given no chance to defend yourself."

"And out of the generosity of your heart, you would like to give me this chance?" Chloe allows the sarcasm free reign. She is not convinced.

"Not exactly. I don't pretend to know what you are going through. I don't know you and I don't owe you anything. You don't owe me anything. I just know that the scientific community is not getting the whole story. They only see what the senior author writes. I think that the other perspective, your side of the story, should be heard as well. And I'm not the only one who thinks this. I have discussed it with my colleagues, and we agree. For the sake of transparency and fair play in science, we need, at least, to hear your version. So it is more a matter of principle than a matter of goodness. And it is up to you, of course. If you don't have anything to add, I can also just go with that. It is completely up to you."

She hesitates, still not sure whether she should talk to this unknown person. But why not? What harm could it possibly do at this point?

"So, what would you like to know?"

"Well, let's start with the most important part. Do you agree with the retraction notice?"

"No, I do not agree with it. I designed this project and all the ideas behind it are mine. I did all of the work, except identifying the chemical inhibitor. I know that the science is correct and the results are solid. I know this because I did it."

"So why is Dr. Palmer retracting it? Surely it is in his best interest to have another Nature article to his name. Why would he want to retract?"

"Well, you will have to ask him that. Unfortunately, he can do what he wants, as the corresponding author. Although it is completely unfair given that it is my work and he had almost no input. I can only imagine that he must be under pressure from somewhere to do this, and he has chosen not to fight. I am being sacrificed instead. That he didn't even have the courtesy to talk to me about it first says a lot. It's all pretty far down on the transparency and fair play scale, to my way of thinking."

"But what about this one experiment that the notice says cannot be repeated? What is your explanation for that?"

"That's simple enough. They just didn't do it properly. What happened was that he, Tom, Dr. Palmer, pressured a number of postdocs from the lab into repeating all my data. A whole bunch of them, and none of them know the techniques as well as I do. Tom doesn't, either. So there they were, unmotivated and in a hurry to finish because it is not their project, the perfect situation for making mistakes. And then we are supposed to believe their results over mine? It makes no sense. The most outrageous thing was that he included the instigator of this whole thing in the farce as well. She probably made sure some of the experiments failed."

"Wow, that does sound very problematic. Who is this instigator? And by instigator, you mean the whistleblower?"

"It's just another postdoc from the lab, no one important. She hasn't published anything from the lab herself and she is obviously jealous of me. I noticed the resentment as soon as my paper got accepted. I just never imagined she would take it this far. Karen is in some ways a sad case. She can't get to the top herself, so instead she tries to pull me down. But the crazy thing is that Tom believes her."

"So she is the person who originally suggested that something might be wrong with your paper?"

"Yes, I'm sure she is. It all makes sense. Tom tried to convince me that one of the mouse technicians pointed it out, as well. But the tech, Andy, would have no reason to mix in. He doesn't understand the work we do, and I doubt very much if he read the paper. He just does his job. Well, he sort of does his job, but not very well. He is in charge of the whole mouse colony, but he does not keep good records. It is a total mess. This is one of the real problems, if you ask me. Poor recordkeeping in the mouse house. Anyway, it has to be her."

"But this person, Karen, was still directly involved in repeating your experiments, even though there was a clear conflict of interest for her? That sounds very improper. How do you know of her involvement?"

"From one of the other postdocs in Tom's lab. He told me a bit about it in the beginning. He told me who was on this silly 'repeat team' and what they had been asked to do. After that, I didn't hear much. I suppose he was warned not to talk to me."

"That sure sounds like a seriously bad setup. So what about the missing original data mentioned in the retraction? I think it said this was part of the reason for the retraction, data missing."

"Some samples were stolen from my fridge, histology slides. Someone took them."

"Someone stole your samples?"

"Yes. That must have been her as well, my accuser. I was away on a trip, job interviews on the West Coast. When I came back, the slides were gone. Everyone knew I would be away for that time. And I have since found out that she was snooping around my desk and my bench while I was away. The fridges aren't locked or anything. So it's not that hard to do. The problem is that I have no proof of this. So they just say that I don't have the data, I am to blame, end of story."

"And that's it? Some slides go missing and the whole paper has to be retracted?"

"Basically, yes. Well, they also wanted me to have kept gel pictures of mouse genotyping data or something like that. But nobody does that, at least not in Tom's lab, nobody keeps pictures of routine genotyping. It's ridiculous. The results were in my lab notebook. I did exactly what everyone else in the lab does and I am in the wrong. It is totally unfair."

"So you are saying that you were following normal practice in Tom Pamler's lab and that, in your case, and only in your case, this led to a retraction?"

"Exactly. I was doing everything the way we normally do things. Nothing unusual. But I suppose I was doing too well, people got jealous and I had to be brought down."

"But still, why would Dr. Palmer, as the head of the lab and the senior author on the paper, go along with this? I find it hard to understand why he would side with your accuser."

"Maybe he was afraid that there would be too much bad press if he stood up for me. Everyone knows what happened to Bolinger. It is safer to denounce someone junior than it is to fight the righteous whistleblowers. If you deny their claims, you may just get yourself targeted. They would be only too happy to bring bigger names down as well. Whistleblowers and their supporters are the modern inquisition. So, I suppose, in some far-out way, it is an understandable choice. Cowardly, absolutely, but then, most people are cowards. It's just totally wrong to hang me out to dry like this. I do not deserve this."

"I completely agree. It shouldn't all be on you. But that seems to be how the system works: find a convenient scapegoat." He pauses and then continues in the same understanding tone, "So are you being formally investigated for misconduct?"

"What? Who said that?"

"Well, Dr. Palmer's secretary said you were on leave, at home. There is a claim that some original data are missing. Your paper has just been retracted, without your cooperation. It is not much of a leap."

"Well, in a way. They are looking into the missing data stuff and I've been helping with that. I am just on temporary leave, until it is sorted."

"And by "they" you mean investigators summoned by the office of scientific integrity at the university?"

"Yes, I suppose so. I mean, that's the way it's done, right?"

"But you are still employed at the Institute?"

"I am. But what does that matter? Can't you see? They retracted my paper without asking me. They ruined my career. They didn't even wait for the investigation to be completed. No one cares about the formal investigation, obviously. It is just a sham. They have decided that someone should take the blame so they can sweep it under the rug and get on with their own precious careers. And that someone is me. Delete, erase, forget."

"And by "they" you are now talking mainly about Tom Palmer?"

"Oh, I don't know. I don't even know why I am talking to you. I probably shouldn't even be talking to you."

"I am very grateful for.."

She cuts him off.

Shit.

That was not the smartest thing to do, Chloe thinks, talking to this guy and letting it all out. She does not even know who he is, this Christopher something. Well, it is just a magazine that no one has ever heard of. Who cares what they do or do not write about this "story"? It is not a story, it is her life, goddammit.

She needs to go for a run, a very long run.

-

She was starting to think that he would never call. That he would conveniently forget he ever knew her. Forget despite his visiting toothbrush and various pieces of clothing left behind. Forget despite the reasonably obvious fact that they have been seeing each other for almost 2 years now. It has been two whole days since the retraction. Two days of solitary frustration.

"Hi—it's Martin."

"Hi there. How are things?"

How come you haven't called is what she really wants to ask. She has considered any number of sarcastic, angry, disappointed remarks for this conversation. But this cool, dissociated voice is what comes out instead.

"Fine, uhm, how are you?" Does he intend this as a real question, she wonders. Seriously? But it seems not. He goes on immediately "I just saw the retraction in Nature. That's very unfortunate."

"Unfortunate? More like a fucking catastrophe." Real anger now.

"Of course, yes. Sorry. It must be awful for you." Neither of them say anything for a while. Martin finally breaks the silence. "But what happened? I don't understand. I thought you said this was all taken care of. You said that Tom knew you had done nothing wrong and that he was fixing it. Why didn't you tell me what was happening?"

"I didn't want to bother you. I thought it might just go away."

"Go away? But the data, what happened with the data? You never told me of any trouble with your data. What did you do?"

"Nothing. Nothing wrong. I told you about the slides that were stolen. I told you. Anyway, I don't want to talk about it any more, OK? I don't. It is all one big screw-up, that's all."

"Screw-up, what do you mean? A retraction doesn't just happen. Is this why you have been staying at home the last couple of months? You said you were writing grants and papers. You acted like it was all settled. But it wasn't, was it?"

"Look, Martin. I don't want to talk about it now. If you really cared, you would have called 2 days ago."

"But I only saw it just now, like I said. And I've been really busy. You know how it is."

"I know how it is. And I'm not stupid. It must have been all over the institute the minute the retraction was online. Prime gossip, who can resist that? You waited 2 days to call."

"Wait a minute. You are the one who has been keeping things from me. You told me that everything was fine, taken care of. Then I read about this huge thing, this retraction, in the journal. Just like that. I heard nothing about this from you. How do you think that makes me feel? You were shutting me out, not the other way around."

"Whatever. Martin, look, you called. That is good. You did your duty. You can go back to work now."

"But Chloe, it's not like that. I really do care."

"Right."

"Chloe, stop it. This is not fair. I just don't understand what has been happening. With you, with your work, with us. I am trying to understand."

"It's complicated. I have a lot to deal with right now."

"OK, OK." He must realize that this is going nowhere. She is sure that he is itching to get away from the phone, from her, and back to his own world. As quickly as possible.

He finally asks. "Is there anything I can do? Anything I can do to help?"

"No, there isn't. Do your thing, Martin. It's OK. Just do that."

"Will you be OK? Will you let me know if there is anything. . ."

"Yes, sure." She cuts him off "Goodbye, Martin".

She puts down the phone slowly, suddenly overwhelmed with sadness. Everything is coming out wrong. She should not be blaming him.

Almost immediately, the phone rings again.

"Martin, I understand . . ." She starts.

"No, it's not . . ." The answer comes, a bit hesitant. "It's Barry."

"Barry. Hi. Sorry. I—ah—I thought it was someone else. I was just on the phone. What a coincidence."

"Coincidence? What? Anyway, I am really sorry for not calling earlier. I was just so surprised. You know, by the retraction. Your lovely paper, suddenly retracted. This is so awful. I'm so sorry. Are you OK?"

"Don't be sorry. I'm OK. Well, no, not really. But it is nice to hear your voice, a friendly voice."

"I know I should have called you sooner. I was just shocked, to be honest. You didn't mention anything about this in all of our long chats. So I've been a tad confused. I thought we had something, you and I, a real connection. And then this pops up, out of nowhere. And you never told me anything about trouble with the paper or the lab."

"Sorry, I should have talked about it. I guess I was just in denial."

"No, I'm sorry." He says quickly. "I shouldn't be thinking of myself. You must be going through something really terrible there. I can't even imagine. But you could have talked to me, you know? No matter what was going on. I'm not out to judge you. That's not what I'm about. It is just, well, the future, and all that. I thought we were making plans. Or was it all just a game?"

"No, no, it wasn't a game. It was absolutely what I wanted, my perfect dream. And it was so tempting to just continue that dream. If only. . . The retraction took me by surprise. There were some issues to sort out, but I thought I would get past it. I really believed I just needed to get out of here. I was looking so much forward to it, to the future."

"All our plans. California dreaming, I suppose. Did you get any job offers before the shit hit the fan?"

"No, I was waiting and waiting for that. I never heard anything. You would have been the first to know. I would have called you in the middle of the night, with that kind of news. Maybe the search committees knew before I did." She only realizes the possibility of this the moment she says it out loud. It becomes a certainty almost immediately. "Tom probably called his friends out there and warned them. That's why nothing happened. It must be. Maybe. No, I don't know. He didn't even talk to me, you know? He contacted the journal to request the retraction. Then, afterwards, he sent me an email telling me what he had done. Fait accompli. No discussion, no respect. My career in an email."

"You know that it's not going to happen now, don't you? There will be no job offers."

"Of course, no one is going to hire someone who just had a Nature article retracted. They wouldn't touch me with a ten-foot pole. I'm sure they are grateful it happened before they got involved with me. Now my CV looks quite a bit less impressive and they have a good excuse to look elsewhere. No one needs to do anything, to say anything."

"What about you? What will you do?"

"I don't know. I just don't know. Another postdoc, perhaps? Second post-docs are not that unusual. I know people who have done really well on their second postdocs and gotten great jobs. Or maybe I should try to patch it up with Tom. He did what he felt he had to do. I could get some other things published and some new things started. I have lots of ideas. I could make my way back up. Get back out there on the job market in a couple of years."

"I hate to say this, Chloe, but it would be incredibly difficult. I'm going to be brutally honest with you. Over time, the ten-foot pole might become just a stick, but it would always be there. It is different from not getting a paper published due to a failed project or being scooped. That you can make up for. But a high profile retraction like this, it stays on record. I don't know what happened with the experiment, but reading between the lines, it looks pretty bad. And it will stay that way. It would be hard to convince a department chair to take a chance on you. And as for making up with Tom and going back to his lab, surely that must be out of the question? I mean, given the recent online article or interview or whatever it is. It seems to be citing you. And it is pretty harsh on Tom."

"The online article, what article?"

"The article in 'At the bench'. I think it calls itself a science magazine but it is more like science gossip. You haven't seen it? The students and postdocs here have been sending the link around. A few of the professors have seen it as well. It makes them very uneasy. Not so much about you, but about their own labs. It makes them worry what a disgruntled postdoc or student might say to a journalist about their lab—things like that. Anyway, I thought you knew about it. The article claims to be based on an interview with you. But, of course, one should not trust these scandal-mongers. He probably just made it up, and pretended that there had been an interview."

"Shit. No, I haven't seen it. But I did talk to a Christopher something from that place. I should have known better. It was a couple of days ago, when the retraction had just come out. I was pretty angry, I guess. But I shouldn't have talked to him. I had never even heard of 'At the bench'. I'm afraid that day I just needed to talk to someone, to anyone, just to get things off my chest. And he seemed sympathetic. Naturally. Sleazy bastard. I have no idea what he has written. I'd better read it."

"I should warn you. It's pretty unpleasant. You know how journalists are. Well, how some journalists are, the worst ones. So Tom hasn't seen it either? He will be pissed."

"I don't know. I haven't heard from him. But I had better look at that article. So I know what I have to deal with. I should do it now. Thanks for calling, Barry. Really. I do appreciate it."

"Of course. Barry, the phone guy, that's me. Take care of yourself, OK? And let me know if I can help with anything."

"Thanks, but think I have to figure this out on my own."

"Still. And let me know if you decide to visit my fair city anyway. You may be a crap scientist, but you are still kind of fun to hang out with."

"Thanks a lot. A crap scientist?"

"Sorry, I can't help it. I often joke when things are serious. You are not a crap scientist, absolutely not. I know that. Quite the contrary. Something went wrong for you. I do not know the details and I don't need to. It can happen, to just about anyone."

"Even scientists"

"Even scientists. It's just that for scientists, the judgment is so much harsher. No second chances. It's black and white."

"Hardly what I need to hear right now. But thanks anyway. Thanks for letting me know and thanks for being straight with me."

They finally ring off, with a vague promise of maybe getting in touch again some time. Maybe.

The article is easy to find. She Googles 'At the bench' and her own name and there it is. It was posted 2 days ago, just hours after the phone call. The title alone tells her where this will be going. She understands why Barry was so sure Tom would be upset.

The tip of the iceberg?
by Christopher Turrell

Earlier today, a high-profile paper published a few months ago in the prestigious journal Nature was retracted. It is one of many such embarrassing recants to hit the scientific community over the past few years. Is everyone becoming more or less honest? It is hard to say. This latest retraction hails from the hitherto well-respected laboratory of Professor Thomas G. Palmer at one of the nations absolute top scientific institutions. What went wrong this time, you may ask. Was it a simple mistake, bad luck or yet another example of scientific misconduct? This reporter decided to investigate. An interview with the first author of the retracted paper revealed shocking truths about how research in this laboratory is done—unprofessionally and without proper regard for scientific accountability. And this is taxpayer-funded research. The laboratory is in effect run by a collection of young, ambitious postdocs who desperately need to get ahead. The busy Professor Palmer gives minimal input and minimal oversight, but is happy to take credit when anyone succeeds. The work from the lab is therefore fraught with problems. Lax procedures and poor recordkeeping seem to be the norm. Add allegations of theft and misconduct and we get an alarming picture of life at the bench in a top laboratory. A laboratory that should be a shining beacon of scientific inquiry turns out to be a den of backstabbing ambition and rushed cover-ups. One must naturally speculate how many other published articles from this world-renowned laboratory and institution are plagued with similar problems. Is this just the first of a wave of retractions? Is it just the tip of the iceberg?

The article goes on with so-called facts, twisted from her discontented words over the phone. Why did she ever talk to this guy? Why did she not just put down the phone? She should have known what he was after. This is not what she said, or not what she meant. But there is always some element she can recognize. Remembering the conversation way too clearly now, she can trace how many of the warped facts of the article link to something she said. Willfully misunderstood and exaggerated, yes, but not totally made up. There is no getting away from this. The interview did happen. Her name is there for all to see. He even fills the reader in on her background. Her (uncontested, unblemished) PhD work, her family background (who cares? How did he find out?). Naturally, he mentions the fact that she was at home, not at the lab. This leads to the ongoing investigation and the alleged misconduct. He even explains, a bit pedantically, the different types of scientific misconduct. Only after this mini-tutorial does he mention that misconduct on her part has yet to be proven, formally. So it is just a formality, by implication. Thank you very much, Mr. know-it-all. He goes through the whole sad tale of accusation and retraction, or rather, his version of it. Karen is also mentioned by name and the possibility of the whistleblowing being driven by intra-lab competitiveness, jealousy and personal vendetta. Once Chloe let her first name slip during their conversation, it must have been easy for him to find out the rest. The efforts of Tom's lab to repeat the results from the paper are ridiculed as a farce, an attempt at cover-up. The reporter laments the inappropriateness of someone who is quite likely the whistleblower being involved in the farce. Come on, Chloe thinks, at least be consistent. Either the repeat effort is a cover-up or it is a jealous vendetta to prove me wrong. It can hardly be both. The article goes beyond the retraction. The author returns to Tom's high-profile lab and how things may or may not be done there. The accusation that Tom does not know or understand much of what goes on in the lab is repeated in different guises. The article rambles along. The smug tone and the relentless insinuations of impropriety are straight out of the tabloids.

After skimming the article the first time, she re-reads it more carefully. She is hurt, of course, but the retraction itself hurt much more. It becomes clear to her that the primary target of this malicious article is, in fact, not her. The target seems to be Tom. The repeated mention of 'lax' procedures and hints that other articles may be incorrect refer specifically to Tom's lab. He is depicted as someone who does not lead and does not contribute intellectually but just takes credit for work once it is done. His prior achievements are mentioned as props, keeping him afloat years after his sell-by date. Much of this malicious speculation about Tom is not based directly on the interview. But the way it is written makes it looks like it is. She has been, and she still is, angry with Tom for how he has treated her. But she knows him. The article makes his lab out to be a particularly appalling example of how science is done in top labs. And it depicts Tom as a useless dinosaur. This, she has to admit, is both untrue and unfair. The article also contains jabs at the famous institute and the 'hypocritical' scientific establishment, but these are feeble and blood-less compared to those directed at Tom. The reporter may say this is an exposé of the world of overfunded biomedical sciences, misdirected (once again) by greedy professors and unchecked minions. But it feels personal. Maybe this Christopher guy has an axe to grind with Tom and her retracted paper is just the means to do that.

These musing are not helping, Chloe knows. The so-called online magazine may just be a blog for discontented scientists with nothing better to do with their time. But the article is disturbingly easy to find and to access. Regardless of the author's dubious motivations, and regardless of where the article is published, it can spread. It already has. That it is based only very loosely on her ill-advised phone conversation is something that only she knows. If enough people read the article and if they take it just a little bit seriously, this could get very messy. Barry said that the link to the article was being circulated amongst the PhD students. As a young faculty member, he would hear from them before the more senior professors. Maybe Tom will never see it. She didn't see it, not until Barry mentioned it. Maybe it will drift away into e-space. She knows this is wishful thinking. She has been wishing for things to go away for a while now, hoping for the world to correct itself to the way it should be. It just doesn't. It gets worse instead.

Gazing out the window, Chloe wonders where this piece of toxic trash was written. It could be from anywhere. Maybe this Christopher guy is a local. Maybe he is sitting a few miles from here, at a university hangout, drooling over the accumulating hits to the article with his friends. That is, if he is computer-savvy and knows how to count the incoming hits. And if he has friends. Maybe he is a former student getting back at Professor Palmer for giving him a bad grade. Or maybe whatever problem he has with Tom is an old wound, opened by seeing the retraction. He did sound like an adult, not a youngster. But voices can be hard to judge, especially over the phone. He could be anyone. Maybe Christopher Turrell isn't even his real name. She could Google him and find out. No, she should not give him the satisfaction of her curiosity. He is not worth it. But the impulse to search the Internet for clues leads her to another thought. The search engine can be used to gauge how much the article is being seen, to some extent. She wonders how closely it is being associated with Tom's name. She types in 'Professor Thomas G. Palmer' and hits return. His page on the institute website still comes out on top, the faculty profile and a couple of key articles below that. But then she sees it, the dreaded 'The tip of the iceberg?' It is prominent, on the first page of the search results. After just 2 days. Tom will be furious.

Chapter 16

The simple blue icon has pride of place on Hiroshi's screen. He credits Skype with saving his life, or at least his sanity. So he has set the program to load automatically whenever he logs on, just in case. He keeps in touch with friends from graduate school who are still in Japan and a couple of them who are doing postdocs in California. And he can deliver regular updates to his parents without them constantly worrying about prices of long distance calls. This means he mostly uses it late at night, when the lab is thinly populated by night owls and far displaced people like himself. It takes Hiroshi a few confused moments to recognize an incoming call alert at midday. It is Akira calling in from San Francisco. Figuring he would not call off-schedule unless it was important, Hiroshi puts on his headphones and clicks accept.

The call is short and leaves Hiroshi puzzled and worried. He opens the email that Akira sent and then the enclosed link. The article is long but the first paragraph says it all. He understands why Akira was concerned. Distaste and reluctance take hold of him, and a touch of alarm as well. He really does not want to read this. But he steadies himself and continues.

Half an hour later, Hiroshi is at the door to Tom's office. It is partially open, so Tom must be in. Hiroshi hesitates. He is tempted to return to his desk, mission aborted. But he does not. Equal parts obligation and the need to unload his burden prevent him from retreating. He knocks on the door and adds a low "Professor Palmer?" while sticking his head in the door.

"Hiroshi, come in. And I thought we had settled this." Tom says with a smile. "It's Tom, please. Everyone says Tom."

"Yes, of course, Tom. I forgot. Sorry." Hiroshi steps inside the office and closes the door gently. He does not sit down, however, when Tom gestures at the chair. Instead he shifts his gaze from Tom to the floor and shuffles his feet, looking distinctly uncomfortable.

"So what is this about, Hiroshi? Some new data for me, I hope. I am always eager for new results." Tom tries a light tone. He likes Hiroshi. He is smart, motivated and has quite interesting ideas. He is also a bit reserved. Sometimes Tom feels he has to coax a discussion of even the most positive of results from him. But it is unusual for Hiroshi to come to his office outside of their pre-arranged meeting times.

"No. It is something else." Hiroshi finally answers. "A friend sent me something. It is not good. I think you must see it. I am sorry."

Hiroshi explains very briefly about Akira, his friend, and that he told him about this online article. "Not nice" is all Hiroshi says about it. He says he will email the link to Tom immediately. He leaves the office before Tom has a chance to get any further explanations.

A few minutes later Tom has the email, the link and the article on his screen. No further explanations are needed. He notices the title, but the byline is what really catches his attention: "Christopher Turrell". This cannot be him, he thinks, knowing at the same time that it must be, of course. The title alone says this will be scathing. The first summary paragraph confirms his suspicions. Not good and not nice are massive understatements. Still, he reads the rest of the article. He reads it quickly, guessing what will come but needing to know for sure. "Tip of the iceberg, you little shit. Aren't you just wallowing in it?" he says out loud, but softly. It has been 5 years since he last heard from Turrell. It was a cryptic note, that time, an email with the title "More trashy science?" and no content. It has been a full 10 years since he last saw him. Turrell tried his best to cause trouble back when he left the lab. Tom was only doing what he had to do. That Turrell tried to fight back was not totally surprising. The later note was more disconcerting, mostly because it indicated that he had not let the whole thing go. But with no desire to consider too deeply what the little shit might think of him, Tom had pushed it aside and more or less forgotten about it. So a public attack like this, after so many years, comes as somewhat of a shock. But, of course, it is a perfect opportunity.

Tom sits, staring at the screen. It is vile stuff and he is furious. He curses Chloe yet another time. This is all because of her and her inability to play by the rules. Damn her. She pulls a fast one and this is what happens. This is what happens to him. The article starts out with the retraction but heads straight for a personal attack on him. How could Chloe do this to him? And why? She gets a rough ride as well. She is made to look dishonest, career-obsessed and basically pitiful. What did she say to Turrell? The article is all malicious suggestions and innuendo. Slander. It is vague enough to make it impossible to prove or disprove. And that is not really the point, either. Tom recognizes this. The point is who has read this trash and who will read it. The formal retraction of the Nature paper was hard. It hurt. And everyone saw it. But at least it was contained, controllable. The impact of this nasty nonsense is quite unpredictable. The retraction will stoke the interest, no doubt. And here is the human-interest side to the story, so-called, with the added spice of a possible larger scandal. How did Turrell get a story online so fast? He must have called Chloe immediately. And then written the piece and posted it at a record pace. But why would he even have seen the retraction so quickly? Does he visit the Nature website every day? Another explanation occurs to him. That Turrell has held on to his anger at Tom for so many years suggests it may have grown to an obsession. Maybe he is keeping track of Tom's publications and activities. Maybe Turrell Googles his name every day. Tom tries it. 'Tip of the iceberg?' comes up on the first page. Not at the top, but still, on the first page. Damn. Pretty soon everyone will see it. Damn you, Chloe, he thinks, again. And damn you, Turrell, you nasty little shit.

Christopher Turrell was an obvious pick as a PhD student. Even back then, when his lab was completely overrun with requests for PhD positions, Christopher was top of the list. Tom usually took two rotation students a year, with the understanding that one of them might stay on as a PhD student. Just might, no promises were made. Now he accepts PhD students only very rarely, although he still gets plenty of applicants. Allison is a bit of an exception. Apart from being smart and hard working, she has a good attitude. What happened that year put him off taking graduate students of the more difficult and moody type. They need so very badly to prove themselves. They need him but they do not want to admit it. And yet, he is responsible for them. They are like pampered, over-intelligent children—and he has had his share. That year, he took on two PhD students. They were so different that it was hard to rank them. And they both seemed quite independent. So accepting both seemed sensible. Christopher was at the top of his class and always had been. Having been an undergraduate at Harvard and now a PhD student here, this was no minor feat. Tom knows the type, not completely unlike his younger self. They are smart, some even brilliant, and do not care to hide it. They are fiercely driven but consider hardworking a derogatory term. Their over-confidence did not used to bother Tom. He could see beyond it. And he used to have a knack for spotting the ones worth keeping. There is the usual thing about inspiration and perspiration. Quite simply, are you willing to put in the effort, day after day? Probably it is essential in any endeavor, he thinks, effort. There are other signs to look for. All thumbs is a problem—experiments have to work. A talent for observation is crucial. And arrogance can negate that. Tom is reminded of Chloe again and how she sparked this whole mess.

Christopher started at the same time as Dan. Dan was a quiet one, not given to promoting himself. But he was tenacious and very thoughtful, given time. He was from the south somewhere, Tom thinks. Dan is his one big regret. Regret that he failed him as a mentor but most of all regret that he did not understand what Dan had found. Dan did not either, but that is beside the point. Tom should have understood. He should have seen it and appreciated its importance. It turned out to be huge. But Tom did not see what was right in front of him, and he knows it. Dan observed some strange dots in cells when they were starved for nutrients and stressed in some other ways. The initial observation was completely serendipitous using an ill-characterized antibody. That remained their best marker, but could not explain what the structure was. The follow-up was confusing. Many other antibodies stained it sometimes, but not always. Dan studied it carefully for a year or so. He collected markers and defined the conditions that induced the little, slightly curved structures. But they had no idea what the structures were for and finally, they let it drop. Dan asked to be transferred to another lab to work on a more well-defined question. Tom was reluctant but had to let him go. A couple of years later, Tom heard of autophagosomes and realized that this was what they had seen. That was a bitter moment. It gave Tom the first inkling that he might no longer be at his very sharpest, though supposedly still in his prime then, in his late forties. In retrospect, he should have encouraged Dan more, helped him get over his frustration, given more ideas. He should have thought of looking by EM. He should have taken the enigmatic antibody staining patterns seriously and not considered the variability due to Dan's inexperience. Instead he let it all slip away and Dan chose another lab. So double regret.

That Dan did not stay in the lab was also due to his interactions with Christopher, Tom is sure of that. Christopher was from the onset convinced that he would come up with something brilliant. Nothing ordinary or obvious for him. He spent endless hours reading and planning. Perhaps Tom should have nudged him on more, explained how doing sometimes leads to thinking, not just the other way around. He thought it was better for Christopher to find out the hard way, by himself, to truly learn. So Tom gave him complete freedom to work on his own ideas, the way he wanted. But as time passed and the brilliant experiments failed to materialize, Christopher started to change. He criticized everyone's work, at group meeting and even worse, insidiously, in private. Everything was judged derivative, or just plain stupid. And he always knew the work well enough to pepper his comments with relevant references. More than one postdoc was reduced to tears. Dan was an easy target. Tom had to make Christopher leave the lab. He was simply too disruptive. In one very frosty conversation, it was all decided. The official explanation would be incompatible working styles. Only Stuart and a few colleagues, including the institute director, knew the whole story. Christopher was less than 2 years into his PhD and with his brilliant academic record, did not have trouble finding another lab, outside the institute, for his thesis work.

He chose one of his other rotation labs, so Tom felt that whoever it was, knew what he was getting. The science was far from Tom's field, so he never heard what happened thereafter.

Well, obviously Turrell's personality did not change, Tom thinks. And he did not forget. That note 5 years ago, and now this. With perfect hindsight and a slightly modified version of the past, he probably feels that he had "understood" what Tom and Dan had not. He had intuitively grasped that the structures Dan saw were a form of cellular trashcans, or recycling stations. He probably forgets that he called lots of things trash back then. Instead, he believes that he alone knew the truth and called it as it was. The famous professor was clueless. That would explain the many hints in the article that Tom does not understand what goes on in his own lab. He is reminding Tom of Dan's work, taunting him. Or maybe is it just pure bile, due to him being rejected? Regardless, this article has the potential to be damaging. Turrell may not have amounted to much (Tom admits not knowing, but considers it a safe bet. Why else would he spend his time picking at old wounds?). But he has an audience now and the Internet at his disposal. This whole thing could effectively be the end of Tom's career, if it takes hold. He could hold on to his job, but what is that worth without his reputation? The suspicions would fester. He is past it, his lab unraveling. The best postdocs would no longer want to come to the lab. Sad decline is not a popular ride. Slow down, a bit less drama, he tells himself. He should not feel so vulnerable, at 57, member of the National Academy, full professor at this prestigious institute. Even if Turrell has an audience now, a year from now, no one will care. They will vaguely remember an unpleasant story, but Tom will still be here, doing good science. That will be the best defense, refuting it all. Except of course his lab will care, right now. If they have not seen the story yet, they will soon. They are also under attack here, in a way. Karen, in particular, is mentioned by name and ridiculed. The others are smeared by the general brush, by the broad insinuations about poor quality work from the lab. He thinks of Hiroshi, standing there just a little while ago, apologetic. Maybe there was a hint of fear in there as well. Hiroshi does not deserve this, either. Tom needs to find out what the best way forward is. What to do with the Chloe situation and with this ugly thing. Two ugly things, they are now. Maybe it would be best to get hold of Sushma again.

"Tom, I got your message and decided to come right over. I had some time and we have a lot to discuss. Deidre assured me it was OK. I brought along Grace Ng from the director's office. I hope you don't mind. Grace, Thomas Palmer."

Sushma's sudden entrance and pro forma apology for taking control is immediately smoothed by the combination of authority and charm that she projects so naturally. Her quick, intelligent eyes seek Tom's and he cannot help but smile, briefly, despite the circumstances. He wants all this to be dealt with as quickly as possible. That Sushma seems as much in control and as unfazed as previously reassures him somewhat. He is more than willing to listen. The woman standing next to Sushma has none of her authoritative poise but displays a quiet professionalism. Probably in her mid-thirties, she has an almost pretty face, thin lips, small nose and chin-length dark hair. A standard gray office suit completes the picture. She holds out her hand to Tom.

"Professor Palmer, I'm very pleased to finally meet you. I've worked on several press releases from your lab." Tom is confused. Press releases? Why is she here? Grace explains: "I handle PR work for the institute, among other things. Usually this means I help translate your discoveries into a language that the public and potential donors will understand. Positive press. Sometimes, like now, there is the potential of negative press. The director's office has to be on top of that as well. We have good press contacts. I am here to help, if I can."

"Right, OK." Sushma steps in, "I thought, given the situation, it would be good for Grace to come along. And I am happy she could find the time." She nods at Grace. Tom clears an extra chair. All three sit at his small round table.

"So, Tom." Sushma pulls out some papers and places them on the table. "I have here the preliminary, still unofficial, report on the investigation into Chloe Varga and her problematic data. Everything we discuss with regard to that is of course confidential. Grace and I have also read the online article by Christopher Turrell. I think we should deal with both of these issues, if we can."

"Yes, absolutely."

"The online article may be what bothers you most right now, but there is very little we can do about it, directly. It is very vague in its accusations. The author is someone you know, I understand?"

"Yes, he was a PhD student with me for a short while, about 10 years ago. He left the lab and I don't know what he did afterwards. We did not part on good terms. I should tell you that. He was extremely disruptive while in my lab and I had to ask him to leave. He made a bit of a fuss, at first, claiming it was unfair that he had to leave. Officially, we called it incompatibility. We both moved on. Or so I thought."

"This explains his antipathy, I suppose. But legally it does not help. He was not expelled from the PhD program? He did not do anything seriously wrong?"

"He did a lot of things seriously wrong. He was mean and malicious to the other students and postdocs. A real menace. He was extremely antagonistic towards the end. But no, I did not catch him tampering with data, if that's what you mean. Nor was he found guilty of any other form of misconduct. I got rid of him before it could get to that."

"I understand. This is just to clarify the situation. I did not find anything specific on him in the files, no charges or official complaints. I just saw that he started in the PhD program, changed labs, but did not get a degree."

"Well, as I said, we were trying to deal with his situation in a way that would not make matters worse. By we, I mean I talked to the director about it and we agreed on a quiet solution. Officially, it was just a transfer of a PhD student to another lab, on the main campus. I wasn't aware that he didn't complete his PhD in the new lab."

"Right, OK." Sushma looks at the papers in front of her. "We will get back to the issue of Christopher Turrell later. But first we should deal with Chloe Varga. We need to discuss how to proceed."

"I have already done the hardest thing. I have retracted the paper. That will forever be associated with my name. Can't we just leave it at that?"

"Well, there was an investigation. Separate from the investigation, you chose to retract the paper. That was your decision."

"Well, yes. I did that because a specific set of data was not reproducible. We did a lot of work to figure that out. And you agreed with this approach, did you not?"

"Yes, sure. We did not oppose the retraction. That is correct. And it was not inconsistent with the findings of the investigation. We did not make you retract the paper, that's all I'm saying. But let's not quibble."

"Of course. You are right. I just want it all to be over with. Immediately. That's a bit optimistic, I know."

"Well, let's see where we end up. Let me summarize the preliminary report on Chloe. Our investigators agree with the statement in the retraction: Raw data supporting Figure 7F could not be recovered. Beyond one set of numbers in a lab notebook, Chloe Varga was unable to produce any evidence that the mice referred to in said panel had existed and had been examined. So we are all on the same page here. Also, in so far as it was investigated, data supporting other parts of the paper were reasonably in order. Data were available and records were consistent. The investigators did note that overall recordkeeping was not optimal. I have looked at their findings and have to agree." Tom starts to say something, but she stops him with a slightly raised hand and continues: "But to be honest, what I see there is no different from what is normal in academic science. I know the standards. So while the general recordkeeping is not immune to challenge, it is not a matter of immediate concern. It may seem like I am going off on a tangent here. But bear with me, there is a reason for considering the more general picture. The quality of the recordkeeping observed could become a serious problem if you were involved in a legal dispute over a patent, for example. That kind of situation could actually cost the university a lot of money."

"Whereas I just lose my reputation."

"Tom, please. We have to focus on the facts here and who can be held responsible. We both know that if anyone had the possibility to do better quality control of the data Chloe put into the paper, it was you. Not to put too fine a point on it, but you chose to believe the graph without checking the underlying data. You put your name on the paper."

"What would you have me do?" Tom almost explodes. "Look at every single piece of raw data before numbers get put into a figure? That's impossible. No one can run a lab like that." He calms himself, so he can explain. "For one thing, the checking would take forever. And my postdocs would see it as a sign that I don't trust them. They are intelligent, ambitious, independent-minded people who have worked years to get their PhD and now have the postdoc phase before they start their own lab. They find it hard enough to accept that I contribute to their projects. They do the work and I 'just' discuss it with them. I help them write the papers, of course, sometimes re-writing the whole damn thing, and I get it published. It is still their project, in their minds. Some of them realize the importance of the discussions we have and the mentoring, but others become resentful when the work is referred to as work from my lab. Anyway, it is a fine balance. If, on top of that, I tried to micromanage everything, it just wouldn't work. No way." He shakes his head for emphasis. "But it is work from my lab and it would amount to very little without my intellectual contributions. And the funding I raise. That is the truth."

"Look, Tom, we are getting off track. We are not here to judge you or to change the way you run your lab. Everyone understands that it is complicated. I was just saying that you cannot have credit without responsibility. That is obvious." He nods in reluctant acceptance. "Now, let's return to the investigation of Chloe and the consequences thereof. You have already retracted the paper, so that is taken care of. Now what about Chloe herself? She is on leave at the moment, as I understand."

"Yes, she has been on leave since the investigation started. That has been best for everyone, I think. Paid leave, though, she has her own fellowship. I think it runs until sometime this summer."

"Right, OK. That's good to know. So the situation is as follows: Raw data are missing but we have no direct evidence of misconduct. Between us, I think we can agree that it is very likely that she fabricated the data. But we have no direct proof of this. That the relevant data are missing is reason enough to retract a paper but not enough to prove misconduct. Positive proof is easier with falsified data, with manipulated or reused images. Here we only have the absence of evidence. We can justify a finding of grossly insufficient recordkeeping, but perhaps not more. Even if it seems highly unlikely, events could have unfolded as Chloe claims they did. There is no evidence to support her counter-claim that material was stolen. But strictly speaking, this too cannot be disproven. Given the other records missing with regard to that one experiment, however, it seems most likely that the theft claim is a defense tactic."

"I agree." Tom adds. "Also, Chloe now says that the whistleblower, Karen, is the person who stole the slides. And Karen has in the meantime proven herself to be fully trustworthy, in my view."

"This is Karen Larsson, who was mentioned in the online gossip piece?" Grace asks. "It said something about her having had an inappropriate role in the further case. What was that?"

"Yes, that's her. She was one of several postdocs from the lab involved in repeating Chloe's experiments. I may have misjudged that. I thought it was for the best. Keeping Karen from doing experiments that were right up her alley might have identified her as the whistleblower. That's what I thought. Maybe I should have kept her out of it. But there was no harm done. The repeat effort was all kosher, I'm completely sure about that."

"The personal angle, fierce competitors at close quarters and a jealousy-driven vendetta, it makes a compelling story. Add to that the suggestion that you tried to cover up with this repeat effort. I know it is wrong, but he has put together some very effective negative publicity." Grace continues, a bit of reluctant admiration creeping in.

"I can see that." Tom says, stiffly. "It is juicy stuff. But what should I do?"

"With regard to the online article, as I said before, I suggest you do nothing. Unfortunately, that is the best advice I can give." Sushma says.

"I have to agree with that." Grace adds, "And in case anyone asks about it, refer them to us. We will prepare a short statement to use from the institute side. We will say that the author of the piece is not an accredited scientist or journalist. The institute knows of his history and his previous hostility toward you. I'll check with the director on that. We will explain that you are a highly respected scientist of long standing and that there is no evidence of any wrongdoing from your side."

"That is not exactly resounding support. No evidence of . . ."

"Well, I understand how you feel." Sushma interrupts "But general statements of support like this, without going too much into detail, can be useful in defusing situations."

"They can." Grace adds.

"So let's get back to the options with regard to Chloe." Sushma continues. "There is no direct evidence of misconduct, and no admission of such from her side. But we do have improper recordkeeping. She is on a temporary contract that would be linked to her fellowship. The easiest option would be for her to stay on leave from the lab until the end of the fellowship this summer. After that, the institute would offer no new contract. Luckily, she is not permanent staff. That would be so much more difficult to deal with."

"That makes sense, I guess." Tom agrees.

"But it does depend on how Chloe reacts, whether she contests it. There are other options we need to look at as well. And by we, I mean the ethics committee. We have our regular meeting on Monday and will discuss the preliminary report on Chloe. We will also have to deal with associated issues, which include the admittedly very loose allegations made by Christopher Turrell. You see, he has also submitted a modified version of the article's text directly to the institute director, as a complaint against you. I can discuss it with you because by posting the article, he made it a public document."

Tom starts to say something, but stops. He should not be surprised. Of course Turrell would try to stir up as much trouble a possible. He signals for Sushma to continue.

"In the committee, we will formulate a recommendation that we pass on to the director. The institute director will have the final word. I am discussing this with you to get your input, but we cannot decide anything in this room." Tom shrugs and nods. "One option is a more in depth investigation of Chloe's previous work. Additional examples of improper data handling could give a stronger misconduct case against her. But this seems unnecessary, given the contract situation. There is yet another option. I know you will not want to hear this, Tom, but it is an option that the committee and the director's office have to consider. Based on Christopher Turrell's accusations, there could be a more expanded investigation of your laboratory. This would cover the work of all scientists in your lab and go back 7 years, the number of years data records are required to be on file."

"You can't be serious." Tom blurts out. "This would completely disrupt the work in my lab. And it would make me look guilty. Don't you see that this is exactly what Turrell wants?"

"Tom, I'm only saying what the options are. I don't think it will come to that. If the director knows the background story on Christopher Turrell, as you say, I'm sure he will take that into consideration. We appreciate that a full lab investigation would be very unpleasant. But you have to understand that the director may feel he needs to protect the institute from allegations of a cover-up. So this option has to be considered. One very pertinent piece of information is which of the broader accusations regarding laboratory procedures come from Christopher Turrell and which come from Chloe. If they support each other and file parallel or joint complaints, then it could become difficult to avoid a full investigation."

"I don't know who is saying what. I haven't talked to either of them. Turrell, I never want to deal with again, if at all possible. Chloe, well, with Chloe, everything was fine until a few months ago. That's why it all seemed so unbelievable. I enjoyed having her in the lab, I did. Recently, I've been angry with her, of course, and we've had very little contact. After this dreadful article, none at all."

"Maybe we need to talk to her and find out what her intentions are, in light of everything we have just discussed. We should do it together. It would be best if I'm present, for obvious reasons."

"I suppose so." Tom hesitates. "It is possible that talking to her could help. I mean, it's not like with Turrell. Yes, I am angry with her. Yes, she has been stupid. But I don't think she is malicious or really wants to harm the lab. Maybe we can talk her around to a sensible solution."

"She might want revenge for what she perceives as unfair treatment. Or she might continue to use an attack on you as a defensive move."

"Maybe, maybe not. She has already lost the paper, so there is not much in it for her now. And honestly, I just don't see her wanting to associate with someone like Turrell. His brand of obsessive vengefulness takes a twisted personality. Maybe I can get her to see some sense, admit defeat and prevent this from spinning totally out of control."

"Possibly. An admission of responsibility from her side could help you a lot. She has not been willing to admit to any wrongdoing so far. But if she does, well, that would simplify things."

"Maybe I should talk to her alone? It might make her less defensive."

"I wouldn't recommend it. Statements have to be witnessed and recorded correctly, if they are to have any value with the committee."

"What if I try to talk to her informally first? I could try to connect with the Chloe I used to know, or thought I knew. If she opens up, I'll make sure she goes to talk to you afterwards."

"That is possible, I suppose. I would need to talk to her directly, preferably in my office. I have to ensure that she is sincere and has no formal complaints, and that nothing is being covered up." This time, Sushma hesitates. "There are no other issues Chloe might want to bring up, are there? If she does not come around?"

"Other issues? What other issues?"

"Any chance she could come up with sexual harassment claims? An attractive female postdoc, stressful work conditions, etc. Believe me, it happens quite a lot. And it could really complicate things."

"No." Tom is adamant. "I have never been involved in anything like that, never. You can look at the records if you want."

"I did." Sushma admits. "Standard procedure. I look at everyone's files when I take on a case. I just wanted to mention the possibility."

Tom is silent.

"OK, then. Let's see what she says. But I'd need an update before the ethics committee meeting on Monday. From you and, if possible, from Chloe." Sushma says as she gathers her papers. Grace follows her lead. Tom gets up as well. His voice is more formal now.

"Thank you both for coming. I appreciate it."

"Of course" Sushma offers, with a firm handshake but no smile.

Exhausted, Tom drops into his chair. One should never assume things cannot get any worse. Sexual harassment? Would Chloe invent something like that to get back at him? No. She is too proud for that, he thinks. That is, assuming he is right that the mudslinging of the article was Turrell's, not Chloe's. It must be. He will fix this, he promises himself. He will not let them shut down his lab. There has been enough damage. He may need to talk to the lab, or at least to Hiroshi, about Turrell's article. Maybe. Maybe not. He is not sure what to say, exactly. But he does need to talk to Chloe.

Chapter 17

Now he wants to talk.

First a dismissive, insulting email, then stony silence and now this. How can they possibly talk now? And what would she say? She simply does not know. Since reading the online article, Chloe has been on a rollercoaster of feelings about the paper, Tom, the lab, her future. At times she is as angry as before and as ready to fight for restitution. This should not have happened to her, it is wrong and unfair. At other times she is uncertain, without foothold, as if she has jumped off a cliff and is still dangling in space. Her unplanned stab at some sort of revenge has spun out of control. She let her guard down and she was used, played by this sympathetic voice over the phone. The article makes her feel sick. How it twists the truth and makes it unrecognizable. It remakes her, redefines her as someone she is not. Knowing that it is being downloaded and poured over with righteous glee by unknown people in unknown labs, more clicks every minute, haunts her constantly. Each click contributes to this new definition of Chloe. If only she could delete it and all its copies. But she can do nothing. Presumably Tom has read it by now. The accusations, the ridiculing, the hatred. And all from her. No doubt about it, he must be furious. Maybe he wants to see her in order to have her officially fired. Based on the interview, she probably could be. She will have to go and find out.

It is almost time to leave. She collects a light coat, walking shoes and a small backpack. It is an unreasonably pretty day and she has decided to walk all the way to the institute. Maybe by the time she gets there she will know what to think.

The building looks as confident, as proud as ever. The red stones, the harmonious curves. She stops for a moment, admiring, trying to connect. But it is no use. The building seems to mock her and her aspirations, her eager ideas, her years of hard work. She cannot pretend otherwise. And there is no point in trying to postpone the inevitable. She is just on time and she should go straight to Tom's office. Her feet know the way.

She steps inside, intending a brisk, straight advance. But she hesitates, instead. After the noise, the wind and the bright sunshine of the city outside, the quiet and the relative darkness inside the building feel oppressive. She has to fight off a strong urge to turn around and leave. Steadying herself near the elevators, she takes off her sunglasses and finds the call button.

Tom is waiting for her. From inside the lab, awkward glances have followed her down the glass corridor, so she enters the office like a shelter. Except it is no shelter, anything but. Tom is on his own, no witnesses or mediators. So it will just be the two of them, trying to have a conversation no one would ever want to have. But at least it is private. Tom finally starts. They are far beyond the need for small talk, so he gets straight to the point.

"I saw the article. 'At the bench'. I understand it is based on an interview with you. As you can imagine, I was shocked by the content, to say the least. Shocked and angry." His voice is firm, controlled, apparently cleansed of the emotions he talks about. "Talking about the retraction with a reporter is one thing. But coming with all those accusations and insinuations, against me, against the lab. How could you?"

"Tom, it wasn't like that. I talked to the guy. I admit that. He called me as soon as the retraction was published with lots of questions. But I did not say the things he wrote. You have to believe me. He twisted my words."

"So you will just stick with denial again, that nothing is your fault?"

"I didn't say that. Of course it is also my fault. I shouldn't have talked to him. But I did not say those things about you. He made up lots of it, I don't know from where."

"But Karen's name, ridiculing our efforts to repeat what was in the paper and so on. That came from you, did it not?"

"Yes, but … Look, I was hurt and angry. Everything was. . ." She pauses, restarts. "We had a good working relationship, before this. We did. I was doing well for myself—and for you. Then suddenly all this happens. You throw the whole investigation business at me without giving me a chance to explain first. You retract the paper without talking to me about it. You threw my career in the trash without a second thought. So I was angry. And yes, I might have said more than I should have to this Christopher character, but I had good reason to be angry. You share some responsibility for that."

"No, I don't." He hesitates. "Well, maybe I could have been more considerate. But look, Chloe, you are not the victim here. This didn't just happen to you. You acted improperly and these are the consequences. You made it happen."

"I did not, I was ..."

"Chloe, give me some respect here. OK? Let's be honest. No more stories. I do not believe in mythical extra mice leaving no trace, or in slides being stolen. I don't believe it now and I didn't then. The someone-out-to-get-you story can't keep covering everything. There is too much missing precisely from that one experiment. Everything except for your word fits with the alternate scenario. The data in Figure 7F were fabricated."

"I am a good scientist. I do not fabricate data."

"You were an excellent scientist, exceptionally talented. That is what makes this so hard to understand. You were one of the most talented postdocs I have ever had in my lab. I mean that. And you had everything going for you, until you made up that result. I believed in you. I was sure you would become someone I would be proud to have trained." Tom speaks quietly, earnestly, looking directly at Chloe. His voice is still controlled and determined, but now with a gentler undercurrent. A potential for sympathy. His elbows rest on the desk; his hands are clasped. He occasionally tries to catch Chloe's gaze, but she remains focused on the bookshelf to Tom's left. "But you've thrown that away now, by taking that short cut. I know you did a ton of hard work. I also believe that you did not fabricate any other data, or do anything else wrong. You are creative, intelligent, tenacious and ambitious. You have all the right qualities to be a great researcher. Except, well, except for humility. I think I understand what happened. You were so sure of how that last experiment would turn out. You had all the other evidence and it was all so convincing that you felt you did not need to do the final experiment. And you did not have the mice you needed. But you were wrong about the experiment, what the outcome would be. We know that now, because we've redone it. Or done it. Twice. Two different people. So we know. You simply could not have gotten the result you claimed you got. You guessed wrong. But guessing wrong is not the problem. Guessing is. You killed your own career by making up data."

He stops, but gets no reply. He waits. He tries again.

"Explain it to me. There must be an explanation."

Chloe shifts in her chair, unclenches the tense shoulders. They wait. Finally Chloe starts softly.

"When I was at Princeton, during my PhD, I worked very, very hard on an important cell death mutant."

Tom nods to show he is listening and will not interrupt. He has no idea where this is going but at least she is talking.

"Starting in the PhD program was strange." She continues, in a flat voice "I had so much experience from my diplom-work in Germany that it was like regressing, being in a classroom again. The other graduate students were children. They had to learn how to be scientists, how to be adults. I already knew. So I was never one of them. Instead I threw myself into my work and worked long hours. It was great. I made a lot of progress. I screened a large bank of mutants and found one with a fascinating phenotype, the opposite of every mutant the lab had been looking at until then. That was CED-14. I worked out the logic of what it must be doing from the genetics and then I found the gene. Seven years ago it was still a lot of work to find the gene when you had a mutant. But I was determined and I succeeded. I found out how it worked at the molecular level, as well. It all made perfect sense, the missing piece in the pathway. It was fantastic." Chloe smiles, remembering. She quickly loses the smile and continues. "I did some cell culture work on the human version of CED-14 as well. Well, you know the story, isn't the one I eventually published in Development."

"Yes of course. That was a beautiful story, I remember. You talked about it when you were interviewing for a postdoc here."

"Well, the key words here are 'eventually' and 'Development'. It was an important finding, but I was scooped by a Nature paper. The worst part was that I had the story ready way before the Nature paper came out. I had even written the manuscript. It was not perfect, I know, but it was a good first draft. Hannah sat on it for months. She never seemed to have time to work on the manuscript. It was always, next week, I promise. She even knew that the Wilson lab was onto something similar, but she never took the threat seriously. Of course, it was much less of a threat to her, at her age, with tenure. The important thing is to do solid work, not rush things, she kept saying. Anyway, you know what happened. Hannah was sorry, of course and I got a nice paper but not a top paper. It was her fault for sitting on it. That is probably why I didn't manage to get a position as an independent fellow here. No top papers. I applied for one around the time I applied to your lab. Anyway ..."

"You didn't want the same thing to happen again."

"No, I couldn't let that happen again. And at first, I was sure it wouldn't. It was great to work with you once I was ready with the story. It was such a relief after Hannah. You gave feedback immediately. I knew that only the experiments, the results, were limiting. We built up a great story and the science was rock solid. But then, last year, there we were. Nature was being incredibly slow. The manuscript had been under review for months and months. The reviewers kept asking for more. Then you told me about the competing group working on JimD10. I got really worried. The reviewers had asked for that mouse experiment. It would have taken 3 months to get enough mice to mate and then another three to do the experiment."

Tom cannot keep quiet. "So you decided to pretend?"

"No, no. It wasn't like that. I didn't just sit down and decide to make it all up. I tried to do it, I really did. But I was so unlucky. I had the two pregnant females, like Andy said. I also tried to do the mating the other way around, but it didn't take, somehow. And then one of the two females died. That was the first piece of bad luck. The other female carried her pups to term and had eight of them. According to standard Mendelian genetics, I should have had two double heterozygotes. I did the tail-clips and the PCR to check. And nothing. I had no pups of the right genotype. None. I just couldn't bear it. It was too unfair, too much bad luck. And, yes, I was sure the experiment was superfluous, anyway. So I did something stupid. I didn't think straight. I made a mistake."

They are both quiet for a while.

"But you could have come to me and discussed it. I could have tried to argue with the editors. The reviewers were very demanding."

"But one month after we had agreed to do the experiment? No, I knew you would say that it was doable. That I would just have to take the time it took. You would have wanted me to see it through."

"I suppose I probably would have. But it would have been OK, you know, the extra time. The competitors were not that close, and they didn't have the inhibitor."

"Well, I didn't know that, did I? I thought they were close. And I. . . I so desperately wanted to go on the job market last fall. I was ready. There was nothing that I wanted more than to finally have my own lab and do my own science. It is all I have ever wanted to do. And it is so important to get the right job, to be at a good place, everyone knows that."

"You could have done that anyway, gone on the job market. With your record and another great paper on the way."

"On the way, but in who knows what journal? No, I would not have gotten interviews at the best places, not without a top paper. I would have had go to Michigan State or some dreadful place like that. I couldn't face yet another so-so publication. Not after all the years of trying so hard. I couldn't have a top paper taken from me one more time."

"But it would not have been a so-so paper. It would still have been a beautiful story even without that last piece of data." He pauses, mentally going through the hypothetical scenario. "I suppose if we had taken the time needed, done the Ras plus Myc experiment and then gotten the negative result we see now, maybe Nature would have said no, but . . ."

Chloe interrupts, "They would have said no, for sure. And then we would have to start all over with another journal. Another year gone by. But I had no idea that the inhibitor didn't work in mice. I was sure it would. It was not like I knew it would be a negative result and I tried to hide it. No. That's much worse. I didn't hide an undesirable result. I was just completely sure I knew what the result would be. I still find it hard to believe that the experiment doesn't work. I know, you told me it was done twice. But I still don't understand."

"Well .." Tom starts. He could say that you have to respect the data, whatever you think they ought to be. That there are no excuses for making up results. That she cannot hide behind not knowing the outcome as a lesser infraction. But there is no need. They sit in silence again. Chloe looks up. She looks directly at Tom and keeps eye contact.

"There, I said it. I told you." She says. "And I am sorry. I truly am. I am sorry that the paper had to be retracted. I know it affects you. But the truth is, it affects me so much more. We both know that my career is over. No one will ever give me another chance. It is different for you. You have so many great papers already. One retraction won't ruin your reputation. Not when someone else is to blame."

"No, perhaps not, but people do care—because I am established. Schadenfreude. People love watching someone well known stumble. And you really stoked that fire by talking to Turrell." He stops himself from going in that direction, from indulging his carefully contained anger. "But why, Chloe, why put everything at risk? Your career, my reputation. It just seems so senseless."

Chloe responds in a voice suddenly bursting with bitterness:

"Everyone is acting like what I did is so goddamn unthinkable. Come on, look around. The reviewers tell us exactly what result we need in order to secure that Nature paper. I cannot believe that I am the only one who has ever succumbed to this temptation. The difference is that I got caught. And it was completely by chance. I work my butt off and do everything right for years and years. Once, just once, I cross the line and then, Bam!, all eyes are on me, everyone is shocked." She catches herself, stops up. "Look, I know it was wrong. I shouldn't have done it." She then leans forward and says, with surprising intensity: "But I just won't buy that everyone else is a saint. I don't believe it."

"Chloe. You cannot absolve yourself by accusing others." He pauses. "There is a line. If you cross it, it is over. You know that. And trying to take others down with you—me, and Karen, as well—that is inexcusable. Why did you say those things to Christopher Turrell?"

"Tom, that part I am really, really sorry about. When I saw what he had written, when I saw how my words had been distorted, I felt awful. I was mad at you, but I didn't intend for it to come out like that. I know it looks that way, but I didn't. Many of the things he wrote, I never said. It's just . . . when he called I was so angry with you. I had just seen the retraction. So I said more than I should have. But I swear, I was not trying to pull you and everyone else down. You have to believe me."

"What you did is wrong, Chloe, and I know you will pay dearly for it. But I also know Christopher Turrell. Well, I knew him, some time ago. He seems to hate me with a vengeance. It's an old story. But it's clear that he used you and the retraction story to get back at me. His motives don't get you off the hook, however. You gave him ammunition."

"And I'm sorry, Tom, I am. I shouldn't have talked to him that day. But this guy, he twisted what I said into something completely different, perverted it."

"He is a twisted guy, a truly malicious person. I hope you see that."

"I do. I hope I never come across him again."

"Well, and I hope my colleagues will recognize malicious slander when they see it. Interest in the piece will fade away, with time. But it's there. And Turrell will certainly try to milk this situation as much as he can." Tom realizes he is straying from his objective, getting too focused on Turrell and his scheming. He still has a few more things to work out with Chloe. "But why did you mention Karen by name? And why did you accuse someone, or specifically Karen, of taking things from you? I find that very hard to accept. You got her name into this slanderous work as well. What made you say that she was the whistleblower?"

"I was sure of it. It was so obvious that she was jealous of me. Obvious from the time we celebrated the acceptance of my paper. Remember? It seems ages ago now. She couldn't even bring herself to congratulate me. Everyone else did, but not her. She just stood there in the corner with a glass of champagne and stared, as if something had been denied her. There is not an ounce of generosity in her. But my paper had absolutely nothing to do with her. We were not competitors or anything like that. After that, it only got worse. She started sneaking around the lab at odd hours. Juan told me that he caught her, one morning, at my desk. This was while I was traveling. She was going through my things, snooping. You can ask Juan. He will tell you what he saw. It is unbelievable, really. So I am sure she is capable of doing something like that, a bit of sabotage, whatever, just to get me in trouble. Just as I am sure she is the whistleblower. It could not have been Andy on his own."

"But she didn't actually take anything, did she?"

"I probably shouldn't have said that, the stuff about the slides. It all just happened so fast. Suddenly there was the accusation, right in my face. You were completely inaccessible, a wall. And you had that self-important lawyer here. I could tell she had already decided what to think. You both had. Afterwards, when I was in the lab with her, I noticed someone had been in my fridge. Things had been moved. So I just said what came to mind, what might have been an explanation."

Tom offers no response.

"So what happens now?" Chloe finally asks.

"Not a lot, I hope."

"Meaning?"

"The paper has already been retracted. That part is finished. And as far as I understand the official investigation of you is clear about inappropriate recordkeeping but is not conclusive with respect to misconduct. So, unless they decide to expand the scope of the investigation, this outcome will be entered into the appropriate files and the case closed. I will inform Sushma Nayar of our conversation. She will need to talk to you directly, in her official capacity. But most likely, we will not need to do anything more public than that."

"And that's it?"

"Well, it's possible that you will have to formally resign. We'll see what they say. I won't insist. Maybe we can arrange it so that your time simply runs out. There's only a couple of months left on your fellowship, anyway. That way you have a bit of time to get on your feet. But you can't come back to the lab, and there will be no extension of any sort. And no recommendations from me. You understand that, of course?"

"Yes. I understand. . . . And my visa?"

Tom looks puzzled. Is she asking him for help? With a formality? Did they not have the same conversation?

"I can't help you with that."

"OK." Chloe pushes the chair back. "I think I will get going then."

Tom stands up as well, searches for what to say.

"Is there anything you want from your desk? Some personal items, perhaps?"

"I think not. Whatever there is, you can throw it out"

On the way out, Chloe walks at a measured pace, looking straight ahead. For the moment, she feels light and free. This will pass, she is sure, and give way to other feelings. But for the moment, this is enough. This time she sees Karen through the glass to the lab. They lock eyes, but Karen quickly looks away. Chloe stands still, to see if she will do anything, look up again or come over. But there is no fight to be had. Chloe moves on. Reaching the elevators, she thinks of the fifth floor. She could pay Martin a visit. Maybe they could try to sort things out. But that chance is long gone, she knows. That life is gone. She pushes the button and looks down the corridor one last time.

Chapter 18

"It actually works. It's amazing."

Three blue-suited researchers are crowded into a small procedure room by the mouse colony. In front of them is a single mouse, on its back, alive and kicking furiously. But it cannot escape Andy's firm grip. It is one of the precious double transgenic mice, the second out of three treated with the drug. It looks exactly like the first one: normal and healthy. Unbelievably healthy. Three pairs of human eyes stare at the mouse, Karen with hard-to-shake disbelief, Lucy with wonder mixed with worry and Andy with calm resignation. Although Karen has more or less known the result for over a week, she still finds it incredible. She reaches forward and feels along the ribcage and abdomen with her gloved finger, first one side, then the other. There are no obvious lumps. The three control mice they examined before these two looked completely different. Their numerous tumors were obvious, visible to the naked eye and palpable to the probing finger. She picks up the camera.

"Could you position it for a few pictures, Andy? Let's do it exactly as we did for the other ones, on its back, legs outwards, pinned down. Thanks." She starts taking pictures, concentrating on the task. When she is done, Andy picks up the squirming mouse by the tail and puts it into the new cage with its already photographed mate. He picks up the last of the three. It looks the same.

"We should talk to Tom." Lucy says. "I mean, he should know that it's working. There's no doubt anymore, is there?"

She looks at Karen.

"No, there's no doubt. And you are right, we have to talk to Tom." Karen says. She can feel her own reluctance. Whether it is making this unexpected result official or the thought of how Tom will react that bothers her, she is not sure.

"We should see him together, all three of us" she continues "In case he has questions." I also do not want to face him alone, she thinks. She knows she has been avoiding talking to Tom about this. But she started it and she will have to see it through. Why did she ever think of this?

It all began a couple of months ago. It was a few weeks after they had had the last meeting of the repeat team. Karen was still sore at not being trusted. She could understand, in principle, that the key experiment had to be repeated more than once. But having Lucy keep quiet about her experiment and thereby using it to test her honesty, that was hurtful. It was hard to let go of this hurt. The others must have wondered about it as well, why her work was being silently shadowed. He should have been grateful that she had come to him with the problem. Instead, he responded by treating her with suspicion. That was uncalled for and demeaning. But at least she got things sorted with

Lucy. Doing that created a bit of a bond between them, something Karen appreciated once other interactions in the lab started to get even more tense.

The online article about the retraction affected everyone. The vague suggestion that other papers from the lab might also be incorrect put anyone with a publication on the defensive. Also, no one had enjoyed being on the repeat team; being accused of contributing to a cover-up attempt added salt to that wound. But for Karen, it was much worse. She was mentioned by name, with insinuations of jealousy, spite and possible misdeeds thrown in. Reading this, she almost cried. She was furious and ashamed at the same time. I did what I had to do, she felt like yelling. But she kept quiet and no one said anything about it, just occasional furtive glances when she passed by. She'd have felt better, she thinks, if they could just get over the awkwardness and talk to her. But they didn't. And now her name was linked to Chloe's fraud. Googling Karen's name, you would now come across that awful article before seeing her real publications. She could not just put it behind her. Just like she could not forget the look on Chloe's face last time she saw her. You did wrong, not me, Karen says to the ghost.

After days of whispered speculations, no one knowing quite what to do or say, Hiroshi finally asked about the article, very hesitantly, at a group meeting. Tom's reply was calm and unemotional. "It is malicious gossip, that's all. We should just ignore it." The answer was sensible, perhaps, but not very useful. How could they ignore it? If he could be cool about it, surely they could as well, was the implication. He was the target, after all. When it was clear that Tom would volunteer nothing more, Juan asked directly about Chloe. Had she left the lab completely, had she resigned? Tom's answer to that was equally brief and dismissive. "I can't discuss that with you. Sorry. All I can say is that she won't be coming back to the lab. I suggest everyone just get on with their own work, OK? Shall we proceed?" Clearly, both the dreadful article and what happened with Chloe were closed for discussion. So they went on with the group meeting. And with time, the lab returned to normal, more or less. Tom had become more distant, though, apparently avoiding nonessential interactions. Karen certainly did not feel like approaching him. There was still too much resentment simmering. If she talked to him directly, she was worried that it might spill over. This would not be good for her, she knew. So when Lucy first came to her to talk about the mice and the inhibitor, they did not involve Tom immediately. It seemed best, at the time. They could tell him later, once they had it figured out.

Lucy had kept the experiment with the inhibitor from Chloe's study going. Soon, all the mice were carrying large tumors and it might be time to put them down. But that was not why Lucy came to talk to Karen. Andy had noticed that Lucy's inhibitor-treated mice were much less active than they should be, almost lethargic. In the control cages, the mice seemed fine, except for the tumors. The tumors were seen in both sets of mice, a consequence of the transgenes. The general well-being problem was only with the drug. Maybe it was toxic in some way. Lucy and Karen went to look at the mice together. They were in the process of changing the bottle with the drugged water when Karen noticed that not much had been taken. Once she noticed this, it didn't take them long to realize that the lethargic behavior might be due to dehydration. Sure enough, when given the choice of fresh, pure water, the drugged mice took to it quickly. And they started to recover. The drug must have smelled or tasted bad, repelled them in some way. Karen realized that not drinking enough could mean that the mice were not getting enough of the drug. If so, then they still did not really know whether the drug worked or not. Maybe the mice drank enough for the drug to be effective. Maybe the inhibitor would have no effect on tumors no matter how it was administered. But it was nagging her. That the inhibitor had no effect in mice when it worked so well in her 3D culture setup had puzzled her, once she got over the slightly perverse satisfaction of seeing the experiment fail so spectacularly. It is not unheard of, of course. Drug metabolism in animals is complex. But it was odd, nonetheless. Low drug intake could be the explanation. She simply had to investigate this further. She suggested they could try to deliver the drug directly, by injection. She would look up related studies and come up with a reasonable dose. The limited communication with Tom meant that none of them had approached him to ask whether the mouse breading should be stopped. So they had plenty mice available, some even mated already. They could easily get enough of the double transgenics to set up an extra small-scale experiment and test the injection idea. So they did.

The small study was straightforward. Andy and Karen identified three plus three age-matched pups of the right genotype. Andy promised to take extra good care of them. Lucy and Karen started the daily injections, with drug or without for the control mice. It meant more weekend-work for Karen and more worries. But she needed to know. Her hopes and expectations kept changing. Now the drug would work, thanks to her reasoning and intervention. Or there would be no effect, like before. If the new approach failed as well, they could lay it to rest. If it succeeded, then what would happen? This buzzed around in her head but never got any clearer. But they had decided to set up the experiment, so they should see it through. Well, she had decided, really. She knows Andy and Lucy would not have gone ahead without consulting Tom if she had not taken the lead. She would have to accept the consequences. And then, last week, the tumors were there. Completely obvious and right on schedule. But, amazingly, only in the control mice. With the drug, no tumors. From then on, it was hard not to check on the mice every day, when they were being injected anyway.

Today is their "official" inspection. Andy brought the camera along, for documentation. The tumors are bigger than last week, but still only detectable in the control mice. The drug works. There is no doubt left. They will have to talk to Tom.

Tom's door is closed.

"He is in, I know." Deidre says when she sees Karen hovering. "He is not in a meeting and not on the phone. So why don't you just knock?"

Karen is not ready. She is shaking slightly. "Lucy and Andy should be there as well. Could you check if he has time this afternoon?"

"Sure, no problem" Deidre looks at her screen. "3 P.M. Is that OK?"

"Yes, sure. I'll let the others know"

Karen retreats quickly and hurries toward the back elevator. No, silly, they are still inside the clean rooms. She turns around and goes to her desk to send emails about the 3 P.M. appointment instead. Four hours from now. Maybe immediately would have been better. How is she supposed to focus on anything with this discussion looming?

Just before 3, Andy and Lucy stop by Karen's desk, as arranged. Together, they go to Tom's office.

"A bit of a crowd today. What's up?" Tom asks, but smiling. Maybe the Chloe-related tension is lifting. He clears space for them at the small table and looks at them in turn as they get settled.

"So, what's this about?" He repeats, when nothing is volunteered. He does not seem impatient, as is often the case, merely curious. Karen realizes that she needs to take the lead here as well.

"We wanted to talk to you about some tumor experiments we have been doing, with the Jmjd10 inhibitor. There were some more MMTV-Ras and Myc mice coming along."

Surprise is evident in Tom's face. "Why are you still working with the Jmjd10 inhibitor? We were done with all that stuff months ago."

"I'll explain that. And I think you'll agree that we did the right thing."

"OK, I'm listening." Tom says, cautiously. He does not look thrilled.

Karen starts from the beginning, from observing that the mice on the drug were not drinking enough. She has decided on a systematic, chronological account, in order not to forget anything.

"They weren't drinking the water? How could you have missed this the first time around?" Tom asks Karen, then turns to Lucy and Andy. "How could all of you possibly have missed that?" He repeats.

"It is not that they don't drink at all." Lucy answers quickly. "They just drink less. It wasn't so obvious."

"But you noticed this now? And not when it mattered?" Tom's voice is raised, unmistakably angry. Karen hadn't counted on this immediate outburst. It throws her off her stride. It clearly upsets Lucy even more. She looks as if she is going to burst into tears.

"Tom, please just hear us out." Karen insists. "Lucy and I both did the original experiment exactly the way it was described in Chloe's paper. Both the dose and the method of administering the drug were exactly as described there. That was what you told us to do."

"But you should have kept your eyes open. You should have been observing. That's what experienced scientists do."

"There was nothing to see back then. We noticed the problem when the mice went beyond 5 weeks. That's when it became obvious." Karen focuses on Tom and adds, slowly. "Anyway, Tom, instead of getting mad at us, you should want to hear the result. The inhibitor works."

Karen tells about administering the inhibitor by injection, the dose and procedure. She opens the laptop she brought along and shows Tom the pictures. He pulls the laptop closer and clicks on some of the images, enlarging them on the screen. He stares at the pictures, cocking his head slightly. Finally, he says, very slowly:

"So, what you are saying is that the inhibitor actually works as it was supposed to? In mice?"

"Yes, it seems quite clear. There is no detectable tumor growth in the treated mice. None. We haven't opened them up yet to look for small nodules, or done histology. We wanted to discuss the timing and so on with you."

"And you are sure that this is simply because you injected the drug instead of putting it in the drinking water?"

"Yes. It's still the same drug, the same batch."

"And you are sure about this? Maybe the tumors are just a bit delayed. Maybe the timing or the age is wrong."

"We treated weaned pups for 5 weeks as before." Lucy jumps in. "We did three of each, matched in pairs, and the controls behaved exactly as before. Only the drug-treated ones are different."

"Of course, we can do it again." Karen adds quickly "To improve the numbers or optimize. We don't mind putting in some more time."

Tom shakes his head. "This is insane. This experiment not working was why I retracted the paper. You both said it did not work. You were so sure. And now the negative result turns out to be wrong? The drug works? This is just too much. I don't know what to believe."

"Last time, Lucy and I both did it precisely as it was described in Chloe's paper." Karen repeats, trying to keep her voice steady. "We did the experiments exactly the same way. The problem is clearly with the experimental design. That's why it didn't work the first time we did it."

"But if I had known this a couple of months ago -" Tom does not finish the sentence. He does not have to. Had he known then, he might not have had to retract the paper. Everything would be different now.

"Maybe you can correct it? Retract the retraction? Once we confirm the result, I mean." Karen says, tentatively.

Tom shakes his head. "You cannot retract a retraction. You cannot go backwards in time. What is done is done."

"But the repeat effort, all the work we did. That was to find the truth, wasn't it? I know it would have been better if we had found out earlier, but we know now. That must still count. It is a good result. I mean, yes, we need more analysis. But the effect is there. We made it work. So the science stands. And the drug works. That must still count. If we get this published then everyone will know the truth."

Karen is aware that she is repeating herself. She thought that Tom would be surprised but ultimately happy to have the original story proven correct. Instead he looks aggrieved and continues to shake his head slowly.

"But why did you do this? We were done with the project. I told you. Karen, this must be your initiative. Why didn't you leave it alone?"

"How can you not want this, Tom? We were trying to help."

"That is not what I meant. Why are you still going after this? It's over. Why does this matter so much to you? I thought you were happy to be done with all this so you could get on with your own project."

"When Lucy saw that the mice weren't drinking, we had to find out if that was the reason the experiment had failed." Karen insists. "We needed, I needed to know."

"OK. Whatever. But there is nothing to be done about it now. We cannot undo the retraction. The retraction notice said that almost all the results were correct. So for anyone who thinks carefully about the science here, that means the logic and all the major findings are still correct. What you have learned now is how to correctly administer the drug to mice. That's all. No more. It does not justify a new publication."

"But . . . but it was so critical just a couple of months ago."

"Yes, it was. But now it no longer is. That's all there is to it."

"But the paper is correct. If you look it up now, you just see that it is retracted. So unless you read the retraction notice very carefully, which no one does, you think it's wrong. If you don't publish this, no one will ever know. They won't know that this inhibitor, a possible cancer drug, actually works. And the lab will never be credited with the findings."

"That's right. But nothing good will come from stirring it up further."

"But this seems so wrong."

"Karen, it is not your concern any more, OK?"

"So what do we do?"

"Nothing. Don't do anything. Just leave it alone. We know about the injection protocol now. That is useful information. I will tell Dr. Singh about it, in case the inhibitor needs to be used for mouse experiments in the future. But we are done with this. It is over."

Karen had imagined many outcomes, but not this, not "leave it alone". She has brought up all the objections she can think of, but they go nowhere. Tom maintains steady eye contact with her. She looks at Lucy and Andy, who look down or away. No one has anything to say. After a short silence, the three of them start to get up. Tom signals to Karen.

"Stay, Karen. I want a word with you."

Karen sits again, with a heavy feeling of dread. She shouldn't have started this. She should have listened to the sensible inner voice of caution. Maybe Tom will berate her for involving Lucy and Andy and that's why he wants them to leave first. She seems to have a talent for doing the wrong thing.

Once the door is closed, Tom starts talking again. But his voice is unexpectedly mild. His irritation seems to have vanished, quicker than she would have thought possible.

"Karen. I think I understand why you feel some responsibility for all of this. But you need to let it go. No one blames you for speaking up about the inconsistencies with the mice, certainly not me. This turnabout is unexpected, for sure. Ironic even, if your mind is inclined that way. But, it is over. I need to let it go. You need let it go. OK?" He pauses, waits for her to nod her acquiescence before he continues.

"Now, I seem to remember your nanotube project was finally shaping up for publication. The results from the cutting experiments that you showed at group meeting last week were totally convincing. It is a beautiful experiment, it really is. And the correlation with tumor progression looks really nice. It gives the work some clear disease relevance, which is good for publication. It is really fine work, Karen. Now, should we talk about getting it wrapped up and submitted? Let's make sure we get it published this year. We can still make it if we work efficiently. Which journal did you have in mind?"

As soon as Karen opens the door, she hears voices from the kitchen. Low voices, gentle laughter and noises of busy pots and pans. The air is fragrant, to say the least, saturated with the smell of curry-leaves and fried chilies. Ashok must be here. She had forgotten about tonight, but with Ashok, it is probably OK. Hopefully Bill will not be grumpy. She steps in with a determined smile.

"Sorry I'm late. I lost track of time."

Bill turns around from the stove with a less-than-content look but no verbal complaint. He gets a kiss. Ashok gets a friendly touch on the arm. It is somewhere in between a handshake and a hug, neither of which seem completely appropriate for him. He smiles back.

"Ashok is teaching me the deep secrets of home-made Indian curries—and pickles, too." Bill explains.

"Hardly" Ashok says, with a hint of regret. "Only the simple tricks that my mother was willing to show me once she realized that she could not send me a wife." Waggling his head and smiling, now in falsetto"Ashok, pay attention now. You will still need to eat."

Karen laughs. Ashok must have fought hard for his bachelorhood in the face of family pressure. He is an interesting man.

"Well, it smells wonderful. And I'm starved. So how were your students this past semester?" She asks Ashok. "I guess exams are over by now? Any budding philosophers of science to set us right?"

"Students are the same as ever, I'm afraid. There were a couple of bright ones this year. They had real spark. But they want to become lawyers or stockbrokers, like the rest of them, maybe doctors."

"Their mothers will approve, then."

"I suppose. So it is pearls before swine, as ever. But I enjoy it. And how is your work coming along? Bill started to tell me that you have been repeating data from an already published paper? It sounded peculiar to me."

"I thought Ashok would be interested in Chloe's story as an example of the strange and messy workings of experimental science." Bill clarifies.

"Well, yes. Actually, I was just finishing that up with Tom today."

"I thought that was months ago." Bill says, halfway a question. "Anyway, it's good that it's over. You had become far too caught up in it."

"Well, it took longer than expected. In the end, having done all the right things just seems to have made everything worse. The whole situation is bizarre and horrible."

"I am afraid you need to back up a bit. I don't understand" Ashok says.

"OK" Karen sighs "You get the ultra-short version, then. A postdoc from the lab published a prominent paper last fall."

"A Nature article." Bill adds.

"And it was retracted because she faked an experiment." Karen continues, "A mouse technician and I found out that something she claimed to have done for the paper could not have been done. The mice she claimed to have used could not have been there. Anyway, I had to tell Tom about it. He told us that there had to be an official investigation. For whatever reason he also decided that we, as a lab, should repeat a lot of experiments from the paper to find out whether the other data in the paper were solid and to repeat the problematic experiment. Basically, to make sure what was published was the truth."

"Sounds reasonable." Ashok says "And was it? The truth, I mean."

"Well, yes and no. Everything was fine, except that last experiment with the nonexistent mice. We did not get the result reported in the paper."

Ashok interrupts her story. "Did you know that a very significant fraction of published results cannot be reproduced? According to the pharmaceutical companies, I should add."

"Yes, I've heard that. That's quite worrying if it is true, isn't it? Anyway, the last experiment did not work at all. We saw no effect when there should have been a strong effect. So Tom decided to retract the paper. It was a big deal. A retraction of a Nature paper looks really bad. For Tom, but in particular for Chloe, given how the retraction was phrased. A career killer. And she's gone now. Maybe resigned, maybe fired, I don't know."

"But was it her fault?"

"Well, yes, I suppose it was. But the point is, I just found out that the result she faked would have been correct, if she'd done the experiment properly. It didn't work when done the way she described it. But I found the flaw and we redid it properly and it worked. So the scientific content of the paper turns out to be completely correct. But now Tom won't do anything about it. He insists that it's over and that I let it be."

"Wait a minute." Bill turns around "What do you mean, the paper is correct? That's not what you told me before—and not what the retraction says."

"I just found out myself. When done correctly, the experiment works. When done as described in the paper, it doesn't."

"So the paper is still wrong?"

"Yes. She faked the experiment, but the conclusion is correct."

"That is complicated, indeed." Ashok looks fascinated. "But this sort of thing must happen fairly often. Something can be technically wrong but the hypothesis proposed is correct. Or it can be technically correct and the theory is wrong. But why do you call it bizarre and horrible?"

"Yes, and why do you care so much? It wasn't your paper." Bill echoes Tom's words from earlier, as well as his irritation.

"Well, because this potential anticancer drug actually works and no one will ever know. The bizarre part is if I hadn't said anything, everything would be fine now, since the conclusion is correct. And it is also kind of horrible that Chloe lost everything. I know that what she did is inexcusable and all that. But it all seems so pointless, so sad."

"But wait a minute" Bill has turned off the frying pan. He is looking straight at Karen, who now looks genuinely miserable. "You said this Chloe was a fraud, a self-promoting, cheating fraud."

"She is that, but she was a good scientist too, before she screwed up. I just feel responsible, somehow. And everyone knows what I did."

"You are not responsible for Chloe's fraud. She is. The only thing that has changed is that you worked out how her experiment should have been done in the first place."

"But no one will ever know. Tom won't publish it."

"That's not the end of the world, you know. I, for one, think you did the right thing all along. She's a cheat, and a liar. She deserves what she got." Bill turns back to the food. Done with this conversation.

Karen wants to say more. She wants to explain what it feels like, walking by Chloe's empty desk. And why she can never meet Juan's gaze. But she can't explain, not that part. She hates not knowing what the others think of her and of what she did. Do they despise her? Do they pity her? She knows she has said enough for one evening. Bill has heard more than he ever wanted to about all of this. That is perfectly clear.

It is Ashok who breaks the silence.

"It sounds complicated regarding this other person. It is not for me to say, of course, but maybe you are being a bit hard on yourself? Anyway, I would much rather hear how your other project is coming along. Last time I was here you told me about those long, thin connections between cells. It sounded extraordinarily interesting. How is that going? Did you get that tricky severing experiment to work?"

How can he possibly remember all that from last year? But she is grateful. Thoughtful Ashok. Eagerly, she starts to answer his questions.

Chapter 19

A dance of life and death—or just idle chatter? What are they up to, these ever-changing little ones? As always, miniscule dramas are enacted under the microscope. And, as always, Karen is there, observing. It is like watching silent films featuring creatures from outer space, she thinks, doing things we do not understand for reasons we cannot intuit. Cells are so interesting. Each one an individual, with so much to do. And, on top of that, even simple cells like these have sophisticated, interconnected lives. Not quite as wired up as neurons, endlessly transmitting information around the brain, processing and remembering. But these little guys are impressive, too, in their own way. She wants to understand how these tiny creatures communicate. More accurately, she wants to figure it out. Learning from a textbook, you get the satisfyingly known, the defined. Learning for yourself, figuring it out, you get the new bits, the unknowns. The best.

One of the cells she is observing is marked with green fluorescence and the other one, some distance away, is red. She can see the now familiar long, thin connection from one cell, green, to the other, red. But this pair of cells has a connection going in the other direction as well, red to green. What does that mean? How does a two-way connection affect cells compared to the one-way connections she has been looking at so far? She does not have the answer, not yet. But it is certainly interesting. She first noticed it about a month ago. It was obvious as soon as she figured out how to make the red cells brighter. Now they are almost as bright as the green cells. Suddenly, there they were, the reverse connections. She has trained herself to recognize them now. Being sure of when they are absent is still difficult, but she is getting there, she thinks. All she has to do now is find out what it means. She smiles to herself—all she has to do—it is a tall order. But she is ready for the challenge. She has a good tool to help her now. The cell cutting experiments have been working better and better ever since she got them to work nine months ago. Progress, but slowly.

She watches the recording one more time, following the two cells until one of the reciprocal connections lets go. Is that normal or an artifact? Have the cells deteriorated from spending too much time under the microscope? She will have to check this systematically. Simply counting the number of connections in samples before and after microscopy sessions should give the answer. She does not think this is an artifact, however. The connections are dynamic. They change as the cells change. Letting go is probably just another aspect of their behavior. But what are they for, she thinks again, why reciprocal? She will have to observe more samples and give it some more thought in order to come up with a hypothesis. This is so exciting. Her hope that finding these thin connections between cells, the "nanotubes" she should say, was just the start of a whole area of investigation seems to be bearing up. There is so much more to learn about them and what they do.

Karen looks at the timer on the screen. It is almost 10 A.M.. In this dark, quiet room it is easy to lose track of time. She has been here for hours and her time is up. As she starts packing up and cleaning up, she continues to ponder the reciprocal connections. She wants to talk to someone about it. But not Tom, not yet. It used to be that she would discuss with Bill, in the evenings or mornings. But they do not do that any more. Part of the truce that no longer seems like such a great idea. She tried to reverse it, to talk casually about her findings of the day. But he seemed uninterested, sometimes even vaguely resentful. So she stopped trying and keeps her ups and downs to herself. But she still wants to tell someone. She wants to share the excitement and speculate wildly about the implications. That phase is always fun, even if proper experiments and reality end up proving you wrong.

Back in the tissue culture room, she sees Etsuko working in the hood she normally uses. When she gets closer, Etsuko notices her and looks momentarily flustered. She makes a movement as if to stand up while still holding a flask in one hand and a pipet in the other, inside the hood. Thus constrained she only manages to move an inch or two.

"No, no" Karen says. "Don't worry, Etsuko. It is not my hood to monopolize every morning. You just do what you need to do. I'll use another one."

Etsuko hesitates, nods a thank you, and settles back to her work.

Karen turns on and wipes down the neighboring hood. With her PhD defense on later today, there is no chance Allison will need the hood this morning. Karen reminds herself to be in time for Allison's talk. There's also the celebration in the lab afterwards to set up. She has to double-check with Lucy that they have everything they need. Even if she and Allison have not talked as much as they used to in this past year, she still considers her a friend. She wants to make sure the PhD defense is celebrated properly. With so much going on this afternoon she had better get moving with the cell culture work now. But she really wants to tell someone about her latest result. Her latest preliminary observation, is more like it. It does not really count as a proper result yet. Etsuko is not too easily excitable, but very thoughtful. Perhaps she is just the right person to bounce this off of.

"I was just looking at my overnight recordings." Karen says "It is so cool . . . I am now completely sure there are reciprocal nanotubes between some of the cells." From behind, Karen sees Etsuko nodding. She is reminded how difficult it is to talk over the noise in here. She will have to wait. "Can I show you later, perhaps? See what you think? If you have time?"

"Yes, it sounds very interesting. Maybe you can show me this afternoon? Or tomorrow?"

Karen considers the busy afternoon, Etsuko's schedule and her own desire for discussion. "Tomorrow would be great. I'll come find you."

After washing up, she starts moving flasks from the incubator to the hood and, one by one, bringing them to microscope for inspection. The door opens and Yuqi steps in. Karen looks up. Something is different about Yuqi these days. The flush and the lingering smile. It is not the slightly nervous, tentative newcomer smile of earlier times. It is different. Maybe she has had some good results as well. Then it hits her. No, of course not, this is something else entirely. She is in love. And Karen has a good guess as to who the lucky guy is. Jeff, the American postdoc who started in the lab earlier this year. Jeff is tall and large, what she would call American size, but at the same time seems very shy. In the beginning, he was always hiding away with his face glued to his computer screen. Recently, however, he has surprised Karen with unsolicited and chirpy "good morning" greetings. The change in demeanor could be due to him finally getting comfortable in the lab and at the institute. But she thinks she has the real reason figured out now. Another lab romance is in the making. The thought makes her happy and a bit nostalgic. She remembers the intense daily turbulence of her lab romance with Bill 5 years ago, when she was a visiting student in Paul's lab. The excitement and complexity of navigating in the lab with him in it, on good days and bad.

As soon as Yuqi is settled in the third hood, Karen tries a gentle teasing.

"You seem very cheerful these days, Yuqi. Any special reason?"

Yuqi shakes her head in an attempt at denial but cannot suppress another notch of the smile. Nor can she suppress the blush from spreading onto her cheeks. But she does not answer directly.

"Jeff seems like a really nice guy." Karen continues.

"Yes. He is very nice." Yuqi blushes deeper but volunteers no more. Karen does not want to embarrass her, so she leaves it at that. She wonders why it has taken her so long to notice it. The two of them may have been dating for months. It is a funny little world they have here.

"So, are you coming to Allison's talk later on?"

"Of course." Yuqi answers, with a slight break in her voice.

"She will do well, I am sure."

"I think so too. I heard that her second paper got accepted"

"Yes, she told me. So it is a double celebration for her."

Karen thinks about her own paper, recently accepted and due out soon. It is a very nice paper, elegant and definitive. She is proud of it, even if it was a bit compromised in novelty. Considering how she thought the whole project was in the dumps when she got scooped back in January, she has done OK. They did not try the most fancy journals, naturally. But she had enough new insight to get it published in her favorite society journal, Journal of Cell Biology. And it did not take much revision to get it accepted. So, she is content. Maybe the next one will be a top-journal story. She will need more than this one paper in order to get a good job. But more is coming. I've just got to keep digging, she thinks. With that reminder, she turns her full concentration to the work in front of her.

"Congratulations, Allison" Karen gives Allison a hug and then steps back to look straight into her face. Allison looks happy and relieved, beaming as she accepts the large bouquet of mixed flowers. "Your talk was great." Karen continues, "And you really nailed those questions."

Karen knows Allison had worried about the talk and especially about the questions afterwards. She had no reason to worry, but such things are not necessarily logical.

"Thanks. It was so . . . fast. It was over before I had time to get really nervous." Allison shakes her head with a short laugh of disbelief. "I just . . ." She does not get a chance to finish the thought as other lab members step up to congratulate her. Karen retreats to the tables behind them and then goes in search of a vase for the flowers. This is Allison's moment.

They are back on the third floor after Allison's PhD thesis defense, which took place in the auditorium. The whole lab is gathered in the coffee area. Karen and Lucy have shopped and decorated, to make it extra festive. There is plenty of champagne and cake for everyone. When Tom returned from the very short closed session, he informed Allison, and everyone nearby, that the committee was impressed and that a celebration was in order. As they knew it would be. No one gets scheduled for an oral defense unless the committee has already decided that the thesis is to be accepted. That would be awful, so why put anyone through it? Despite the pro forma nature of it, Karen likes the tradition of an official PhD defense. It is an opportunity to celebrate someone officially joining the ranks of scientists. She thinks the American way is a bit lightweight, though. But it is still an event.

The PhD defense events Karen has attended have all included public talks. Parents and friends come along for the occasion, whether they understand the science or not. Allison's parents were at the talk today, as well. Who else could the middle-aged couple in the second row have been? Karen sees that they are here for the lab celebration, as well, standing behind Allison, near the door. Like Allison, they look dressed up for the occasion. They hold hands and keep their eyes on Allison. How nice, Karen thinks, their daughter getting a PhD degree probably means a lot to them. Maybe they are also curious to see where she has been working all these years. They move into the crowded room and look around tentatively. With Allison busy and no one else to catch their attention, they end up looking through the glass walls into the lab, its benches filed with the paraphernalia of science. Karen wonders if they have ever seen a lab before. Tom notices them and steps over to chat. He will be trying to make them feel welcome and at ease. The direct attention from the famous professor obviously pleases the parents and they quickly settle into conversation with him. They will remember that. Karen watches and smiles, recalling her own PhD defense. The same thing happened. Her slightly overwhelmed but proud parents were given the attentions of the gracious older professor. It was a different country and seems so far away. Funny, the patterns that repeat, she thinks. Soon it will be time to pop the champagne and then a short speech from Tom.

It is the first time the lab has had a celebration since Chloe's paper, now a year ago. Karen acknowledges to herself that she actively avoided celebrating her paper some weeks ago. Tom seemed content with this, she sensed. She could not tell whether that was because of something with her, with the paper being in JCB, or more of an unwelcome shadow of last year. A PhD defense is a different thing, though. Everyone is happy to celebrate this. She looks around. Despite the normal turnover in Tom's lab, it is almost the same group as a year ago. Jeff is new. And Chloe is missing, of course. And Juan. That was a strange story. A sudden job offer from Madrid and he was gone in a flash over the summer. There was talk that perhaps his family had not settled in well and wanted to go back to Spain as quickly as possible. This job vacancy had come along out of the blue. He told people that it was his dream job, so it would be crazy not to apply for it, even if it was early in his postdoc time. And he got it. She is not sure this is the whole story. He might have had other reasons for wanting to leave the lab. Maybe he felt as awkward about everything as she did. For Karen, not seeing Juan every day has made it easier in the lab. Days now go by without her consciously thinking about all that. The weight of regret and self-recrimination is slowly lifting off of her shoulders. But sometimes it comes back. Like right now, when the whole lab is assembled, waiting for Tom to give his little speech of congratulation. She shudders. She does not want to think about Chloe now. The only way to avoid it is to do something. She sees Vikram close by, unengaged at the moment. This is a good opportunity to ask him about his recent paper. She heard that it has just been accepted for publication, so he will probably be happy to talk about it. Before an appropriate question about his paper presents itself, she remembers today's occasion and uses that for an opening, instead.

"It is so rare that we have a PhD defense to celebrate in the lab. We are almost all postdocs. I'd forgotten how nice it is."

"Yes, everyone is all aflutter." Vikram rolls his eyes and makes slight flapping motions with his hands. "The parents always seem so out of place. They are dreadfully proud. But they are not sure of what, exactly." He pauses, but then continues in a tone of exaggerated resignation. "Of course, in these gloriously democratic days, getting a PhD in molecular biology is not exactly a towering accomplishment. It requires 5 years of toiling at the bench and your professor's willingness to put your name first on a paper. Even if said professor wrote the paper and thought of all the experiments. Contributing with significant scientific thoughts is entirely optional. But we don't need to tell the parents that."

"That is a bit harsh, isn't it? Allison knows her stuff."

"Oh, not Allison, of course not. She is a bright little thing. It was just a general commentary on the sad state of affairs in our world. You have to admit that I am right, no?"

"Well, yes, in a way. I know this happens. But some PhDs are very deserving, and I think we can count Allison among those."

"Yes, yes, she deserves it, of course. But I've been around, you know", with a twinkle in his eye, "I have seen the riff-raff with PhDs who dwell outside Route 128." Now more serious, "It is a bit depressing. The highest degree awarded by the University—the PhD—and it goes to well-behaved technicians. And the defense is just a formality."

"And your PhD defense was not?"

"Well, I was put to the test by some very ferocious examiners."

"So it was not like today's event then? That's funny. I was just thinking how unexpectedly similar this was to my defense. Well, in some ways, at least. In Sweden, we have an official opponent, a professor of the appropriate expertise brought in to challenge you, on stage, in front of everyone. They are required to ask more questions, so it generally drags out for a bit longer. But it is a formality as well. Heads do not roll. The champagne is on ice."

"Well, my Viva, that is, my PhD defense, was different. In the UK, these things are meant to bring out a bit of a sweat. Trial by fire, and all that. I had hours of arduous grilling in my subject plus endless questions in more remote areas of science. We even managed some nonsuperficial intellectual discussions, like real scientists."

"But you passed, it seems."

"I passed, yes." Vikram flashes a playful grin. "But then, as you may have observed by now, I am also extraordinarily talented. So naturally I passed my Viva with excellence. I impressed the examiners so much that they asked me never to talk about it."

This is means to elicit a laugh and it works.

"So, very talented Dr. Vikram, what is this I hear about your massive tumor heterogeneity study? Has it been accepted in Science?"

"Not quite, but it will be. The reviewers just asked for some more data analysis and statistics. Fortunately, these are things I can easily do. Thank Gods—all of them," he smiles."There were no requests for additional experiments. So next month it will be a done deal, I hope."

"That's wonderful. Congratulations. When you last talked about it, I was so amazed at the scope of the analysis. It seemed like 'omics squared. It must also be a great starting point for future work. So will you be going on the job market soon?"

"I most definitely will. I need to move on and stand fully on my own two feet before I get too long in the tooth. I see a professorship calling. But I will settle for an assistant professorship."

"Long in the tooth, you? With a PhD from the UK, you are probably still in your twenties."

"No, my dear, I have hit thirty. Just a couple of months ago, but I am afraid my best days are over. Can't you tell? Plus, I have an old soul."

"Thirty, that's nothing. I am already 3 years your senior and not anywhere near done with my postdoc. I will be old in body and soul before I get a real job."

They are both silent for a moment.

"That was so unfortunate, you getting scooped. A real shame." He says, with genuine sympathy now. "You were just incredibly unlucky. It would have been great if you had been first with nanotubes. But your JCB paper is a paper to be proud of." He adds "And I am not being condescending here. I mean it. It is beautiful work. Just bad timing."

Karen is quite touched by Vikram's words. He remembers what happened to her and is gracious enough to turn it into a compliment. But she does not really want to talk about herself and her career prospects now. A thought that occurred to her when she first heard about Vikram's paper returns to her now and she blurts it out.

"I suppose all these papers getting accepted must mean that Tom's reputation was not seriously affected by the retraction of Chloe's paper. I mean, your paper on its way in Science proves that. Our projects are unrelated to the Nature paper, of course. But if the community had started mistrusting Tom, it wouldn't all move on so smoothly."

"I agree. Tom has managed this quite well. It is just a little glitch on his trajectory. He won't be seriously affected. I am not sure whether I think this is totally fair. But I have to admit it is a relief for the rest of us. If he had been ostracized by it, we would all be in trouble."

"That's true."

"And that, my dear, would be ever so slightly annoying." Vikram tries to return to light banter. But Karen does not go along this time.

"What about Chloe? Do you know what happened to her?"

"No. I have no idea. She was here one day, gone the next."

"It worries me, sometimes, that no one has heard anything."

"That is just the way it is. Everyone has his or her own stuff to worry about. You must remember that we are not family. The lab is just a temporary harbor, like sharing a flat. It can be enjoyable, but don't mistake lab camaraderie for real closeness." Vikram finishes with an unexpected edge. He must have had his share of disappointments.

"I was just thinking about her, that's all. The celebration reminded me." They look over to Tom who is making moves to start his congratulatory speech to Allison.

"I understand." Vikram says, more gently. They hush to let Tom speak.

Tom's standard words of congratulations and his usual gently self-deprecating jokes do not demand much attention. Karen's thoughts go elsewhere. She thinks about her disastrous visit to Martin's lab some months back. She was desperate to know what had happened to Chloe. Although still angry with her for the things that she got that reporter to write, Karen's complex feelings of responsibility were even stronger. Tom was not saying anything. He acted as if Chloe had never been there. No one else said anything either. They probably did not know much, but still. Moving on, everyone was moving on. What if Chloe was in a deep depression somewhere and no one noticed? Like most of them, she would be far from family, relying on friends and colleagues. Finally, Karen gathered all her courage and went to speak to Martin. It was her first trip to the fifth floor. And her last. One look at him as she entered his tiny office and she could see that coming there was a mistake.

"It is just . . . none of us in the lab have heard from Chloe." She tried, anyway. "Do you know where she is . . . I mean, do you know if she is OK?"

"You've got a lot of nerve. What makes you think I would talk to you? And especially about Chloe?" Martin practically spat back. "Haven't you done enough? You ruined her career, you know? And she is a much better scientist than you will ever be. Go away."

Karen remembers stepping backward from the force of Martin's angry words. She could find nothing to say. No defense, no justification came quickly enough to her. Instead she just retreated as fast as she could out of the office and down the hall. Ever since, she has made sure to avoid Martin and the fifth floor. Luckily, that has not been difficult. But his every word is firmly written in her memory.

Lively chatter and Allison making her rounds with a glass of champagne return Karen abruptly to the present and to the happy occasion. Noticing Karen empty-handed, Allison picks up a recently filled glass and hands it over.

"Toast with me," she says and they do.

"So, are you enjoying your big day? And your parents?" Karen pushes away the unbidden recollections and mounts a reasonably genuine smile.

"Absolutely. I'm so relieved it is all done. And all this, it is so nice. Thanks so much for doing it, you and Lucy."

"No problem. You deserve it, after all the hard work." Karen sips some of the fizzy drink. "So what are your plans? Are you coming back to the lab for while or moving straight on?"

"I'm moving on. I have a bit of time with my family and then I am off to start my postdoc. I have finished up what I needed to do here and. . ."

"And there are better things waiting out there?"

"Exactly." Allison says with a giggle. "I can't wait. Almost a year of very long-distance with Alex is quite enough. But we will stay in touch, right? Or maybe we will meet at a conference somewhere?"

"Sure." Karen says, smiling. With so many goodbyes, occasionally the "see you" must come true.

Chapter 20

December 2014

The lights in the conference hall switch suddenly from the bright artifice of activity to somber, after-hours dim, rousing her. She must have been far away. The stingy illumination reminds the last stragglers to pack up and go. Rows upon rows of poster boards and the meeting's many banners are being stripped away. Soon the hall will be taken over by another large group—discussing trends in house sales or electronics, who knows. Some years back, Karen would have dismissed these efforts as trivial. Now she is less eager to judge. But she remains content that her lot is in science. Despite the tough times, the stumbles and the long hours. Despite knowing that that each advance is almost ridiculously tiny, both her contributions and those expertly spun by more famous practitioners and used to mark off precious territory. It is her world and she has made her small place in it. Today was proof.

Leaving the conference hall behind, she is hit by clear afternoon sunlight in the glassy foyer. She heads for the elevators leading to the hotel floors, to her room. With her talk being on the last day, she decided not to rush the travel home. She now thinks this might have been a bad idea. Why should she stay in Boston for an extra night? She did what she came for. She gave a good talk. Seeing Torsten today after all these years was fun, though. And Allison, of course, it was nice to see her yesterday and to catch up. No one else from Tom's lab was at the meeting, it seems. They must be scattered all over by now. Those imagined sightings of Chloe seem crazy now, just nerves. Why would Chloe even be in Boston? Or show up here? She wouldn't. Karen presses the button for the 17th floor. She will resist the trap of logging onto her email to see what minor crises might have befallen the lab in the last 24 hours. Instead, she will go for a walk in the nice weather, play tourist in Cambridge. She deserves to enjoy a job well done. Too bad that Tom was not here. But then he never goes to these big meetings, not unless he is giving a plenary or keynote lecture. Understandable. Still, it would have been nice if he had witnessed her success. The institute is only a few miles away, after all.

An hour later, Karen is walking past Harvard square, watching the crowd. The sunny afternoon has filled the streets around the campus. Students, she presumes, sporting backpacks, colorful knit caps and frayed jeans or tights. But the cold winter air and waning light soon draws many of them, like her, to the cafés lining the street. The one she enters is warm and cozy and the smell of coffee is heavenly.

A few of the students in the café are talking or reading. Most are fiddling with their phones, or tablets, intent on the lit screens. She wonders idly how the cafés make money, with most chairs taken up by study-hall-transplants. She feels no nostalgic yearning to be in their place. The setting is too different from what it was like in her own student days in Stockholm. Café visits were not common then. Nor were they, the students, always connected to the nebulous e-world. There were long stretches of actual solitary time, scary or comforting, depending on your personality. It is strange how just a decade and a half can make you feel so outdated. These kids live in a different world. But the principal reason for her lack of nostalgia is that she remembers. Those were not glorious, carefree days for her. She worked hard, to the exclusion of many other things. She is now reaping the benefit, having given a well-received talk at the ASCB meeting. She feels content, satisfyingly accomplished, as she waits to put in her order for a cappuccino and a piece of carrot cake. A small indulgence for a special day. Warming up quickly, she takes off her solid Chicago-proof winter coat and drapes it over one arm.

The tables by the windows are all taken, so she moves to the rear of the room to find a place to sit. One face there looks familiar. Chinese. She is talking to a man that Karen can only see the back of. Yuqi, of course, that is who it is. And the man could be Jeff. Karen moves in their direction and starts smiling even before Yuqi looks up and recognizes her.

"Karen, It's you. So good to see you." Yuqi says, returning a smile of her own. "Great talk, this morning. You remember Jeff?"

Jeff turns halfway around and, seeing her tray, gets up to help.

"Thanks" Karen says, to both of them.

"Sit, sit." Yuqi clears some space and pulls over a chair.

"So, you were at the meeting?" Karen says and sits down.

"Yes, we were." Yuqi answers. "What a huge meeting. We were just talking about how easily you get saturated. At least we do, no longer fresh students. And after four days, for sure. But it was fun. We wanted to say hello to you after your talk, but there was such a crowd around you. I'm so glad you found us here."

"Did you see Allison?" Karen asks. "I talked to her yesterday at her poster. First time I've seen her since she left the lab."

"No, I didn't. How is she?"

"Very well. She showed me pictures of the cutest kids. One and four, I think. She was beaming like mad when she talked about them, proud and full of joy. I got the feeling that she was having a tough time juggling family and work, though. The usual problem. But she seemed happy. And her work is going OK. She's finally getting things published from her postdoc. But, tell me, how are you? And where are you?" Karen adds quickly, looking from Yuqi to Jeff and back again.

"I moved to a small Biotech startup, Genelocks, about a year ago." Jeff says. "Here in Boston. It's very different from being in Tom's lab, but I enjoy it. It was the right choice for me. I just hope the company stays afloat for long enough for me to do something useful. Or that I at least get enough industry experience for the next move to be upwards."

"And what about you, Yuqi?"

"I ended up staying in Tom's lab. As a staff scientist."

"That's new. New for Tom, I mean."

"Yes. He felt he needed someone to keep core projects running in the lab, the Myc-related things. I have undergrads and an occasional rotation student to help out. Tom has been good about giving me flexibility and some independence. I do lab management stuff, as well. A bit tedious, but it's OK. Deidre helps with that part."

"So Lucy is no longer lab manager? I thought she took care of that side of things."

Yuqi hesitates, looks away briefly.

"You haven't heard?"

"Heard what?"

"About Lucy. She was diagnosed with pancreatic cancer about 4 years ago. She went downhill pretty quickly. It was really tough. She died about a year afterwards."

"My God. No, I had no idea. That's terrible. Poor Lucy. She was such a nice person. And she was only in her forties, wasn't she?"

"Close to fifty, I think. It was quite a shock for Tom."

"I haven't been in touch since I left, I have to admit. Just a few emails early on. Nothing since." Karen says.

"You should go see him." Jeff says, with sudden certainty. "Tom likes to get news from his 'offspring'. I really enjoyed your talk this morning and I'm sure he'd love to hear about it."

Karen responds with a reflex "Thanks" before Yuqi chimes in.

"Yes. That's an excellent idea. Tom should still be in his office for another couple of hours. He usually doesn't go home until 7 P.M., after the traffic."

Still rattled by the news about Lucy, Karen does not manage to find an answer immediately.

After a moment of quiet, Yuqi and Jeff pick up the conversation again. Taking turns, they go through former lab members Karen might know and explain where they are now. Apart from Allison, Karen has heard from no one, she realizes. Yuqi and Jeff seem happy to be able to update her. She asks occasional questions as they move along the list of lab alumni. Most are success stories, of a sort, like Karen's. Assistant professorship here or there, a few other choices made. Vikram is doing very well, no surprise there. Karen volunteers a bit about her lab, her students, life as a PI, how much she enjoys having her own lab. But she avoids her private life. She was never really close with Yuqi, and not at all with Jeff. So they do not ask. Allison did, of course. Karen had to tell her about Bill. How they did not make it through what she now knows is a tough time for many marriages: The assistant professorship years. She had promised Bill it would get better, after the years of long hours in Tom's lab. She would have more time, soon. But soon never arrived. Instead, she got busier, caught up with starting her lab and getting things done, papers published. Luckily he was just on leave from his job in Boston, so he could get it back. Bill is OK, now, she thinks. And she is OK.

Karen realizes she has finished her coffee and her cake. It is time to move on, before it gets too cold outside. Yuqi and Jeff seem content to stay where they are. So she says her goodbyes. It was nice to catch up on everyone's whereabouts. But Karen's thoughts keep returning to Lucy. She has not been in touch with her, or even thought about her, for these past 5 years. She did not really know Lucy all that well. Still, it feels so strange that she is gone, and has been, for years. Completely, absolutely gone. The simple fact, it is shocking, somehow, she thinks, unreasonably so.

Minutes later, she is walking at a brisk pace down Mass Avenue. She knows where she is going. Of course, this is what she should do. Tom will probably still be in his office.

At the entrance, she has to be registered as a guest with security. Deidre has already left for the day, so they call Tom. It will not be a complete surprise when she knocks on his door, then. That's OK, she thinks. And I will not stay long.

The third floor corridor is unchanged, with its generous view of the busy labs. But feels so different. She suppresses a sudden desire to leave. It feels like trespassing. There is not a single person that she recognizes. Yet they all seem to be at home here, comfortable. Of course they are, they belong here now, these new postdocs and students. She takes note of the labs still being busy, although it is almost 6 P.M. People are doing experiments or tapping away on their keyboards, fully engrossed in their work. No one notices her passing by. She wishes her lab were like that. Busy into the evening, full of driven students and postdocs. Stop, do not complain, she reminds herself. With limited drive and few ideas, her students may never become serious scientists. But they are nice kids and they get things done. As long as she tells them what to do, that is. Not like herself as a PhD student. Anyway, she has adjusted her expectations, adapted and accepted. At least she thinks she has.

Tom has left the door ajar and looks up with a warm smile when she enters the office. He gets up and takes a few steps toward her. As he moves free of the desk, she notices that he has aged. His face is only slightly changed, a bit thinner perhaps. And the cropped hair may be marginally grayer. But he seems less tall somehow, less robust, and carries subtle hints of future frailty. He is not the intimidating presence she remembers.

"Karen. What a pleasure. I heard you gave a fantastic talk at the ASCB meeting this morning. Congratulations."

"Thanks, Tom. I am very satisfied with how it went; relieved, too. And it was a good turnout."

"And good papers, I've noticed. I'm very proud of you. Of course I always knew you would do well." He says with what seems like a wink. Maybe this refers to the 'always knew' part. Or maybe she is just imagining it.

Happy that he has noticed her publications, Karen tells Tom about her lab and the projects they are pursuing. Focusing on the positive. And why not? Tom seems interested, asks questions along the way. As they return to the meeting and her successful talk, she remembers to mention Allison, her cute kids and her career prospects. Tom nods along. He has probably heard from her directly. Allison is someone who would stay in touch with her former mentor, she realizes. Not like Karen. She also mentions meeting Yuqi and Jeff. This gives her a chance to say how sorry she is to hear about Lucy.

"It was a bad turn." Tom says, quietly. "She fought really hard and tried everything the doctors could think of. But it is pretty much an impossible cancer to beat. She was really frail by the end." He looks saddened, remembering. "It seems so wrong. So wrong. If we ever need a reminder of why we do the research we do. . ." He trails off.

"She was a really nice person. And she really cared about the lab." is all Karen can think to say. A bit shallow, she fears.

After a few moments of silence, Karen asks about new work from Tom's lab. She has kept up with the published work, so she can ask the right leading questions. This enlivens him and he tells her about a couple of stories that are to be published soon. He is still doing interesting science. This reassures her, somehow.

It is nearing 7 P.M. and they are winding down. Karen realizes there is one more thing that she wants to ask about, but she is unsure how to approach it. Unlike the rest of today's conversations, there is no template for this, no agreed upon level of sharing. She simply asks.

"Tom, I've been wondering. What happened to Chloe? I mean after."

"What do you mean, happened after?"

"Well, you know I felt bad about the whole thing with her."

"You shouldn't. You did the right thing. She did wrong. And she admitted it to me, finally. So it could be put away."

"She did? She admitted it? I never knew. The whole thing was so murky at the end."

"There was something strange between you and Chloe." Tom suddenly sounds very tired. "Wasn't there? It is not unusual, rivalry among postdocs. But you shouldn't hang on to it. It does no good."

"Maybe I did obsess a bit at the time. It was a stressful time." She pauses, but decides to continue. "But what did happen to Chloe? I Googled her later on—just Chloe Varga—and never found anything after—after all the mess, I mean. So I sort of worried. She could of course have gone back to Europe—but still. . ."

"No, she is still in the US, I think. She landed on her feet. Got a good job consulting with McKinsey and moved on from there. She had—has, I suppose—lots of talents. And she was smart enough not to let what happened bury her completely. To move on."

"But how come I couldn't find her, then?"

"By searching online, you mean? Well, she changed her name. Married and changed her last name. Quite sensible, under the circumstances."

"Of course. Makes sense. I don't know why I didn't think of that. So you are in touch with her?"

"No, not really. We're not exactly on friendly terms, you understand. But she has sent me a note every now and then."

So no big drama. Just life.

One last question. Karen has to ask. "Did you ever tell her that the drug actually worked? That it worked in mice, I mean?"

"No." Tom looks at Karen, puzzled. He shakes his head, whether for emphasis or in disbelief, it is hard to tell. "No, we haven't talked much. As I said."

It is time to leave. Karen starts thinking of the walk back, or a bus, for old times sake. Something to eat, then sleep, an early flight tomorrow. They say their goodbyes. Maybe they will meet again. It is a small world after all. And she is somebody now. Not famous, but in the game. She remembers to tell Tom how much she appreciated being in his lab, how much she learned and how happy she was there.

As she leaves Tom's office, her last words still resonate. Happy? Who knows? Maybe she was.

Part II

Questions and Answers with the Author

Q and A With Pernille Rørth, Author of Raw Data

1. Why does your story about life in science include a case of misconduct? Is it directly inspired by specific cases?

This is a story about life as a scientist, the thrills as well as the hardships. It is focused on two young scientists at a critical time of their careers. They are ready, or almost ready, to compete for their first faculty position and step fully, independently onto the science scene. This is a stressful time for most scientists. The intensity makes it an interesting career-stage to depict. Scientific misconduct, cheating, was included in part because of the popular interest in the subject and in part because, for scientists, misconduct is a big deal. The high stakes, plus the need for investigation, also makes it a useful plot device. The story is, however, pure fiction and not a fictionalized version of a real case.

Media interest in misconduct is understandable. Most scientists are funded by public money, so they should serve the public good. They are generally respected by society and vested with a fair amount of trust. When this trust is broken, people are disappointed and upset. They want to understand why such misdeeds happen and they want to see action taken to prevent it from happening again. Funders and institutional stakeholders want to better understand the roots of misconduct as a step toward prevention. Of course, there are also less salubrious reasons for a popular interest in the subject. It seems to be human nature to be drawn to the failing and fall of others—the more spectacular, the better. Stories of major misconduct or fraud can be attractive from that angle. If the accused party is a strong and interesting personality, perhaps defiant, and this person fights back, this adds spice to the story. Some of the more dramatic recent cases of this sort in biomedical science are the bizarre Penkova case in the Danish media and the STAP cell case in Japan, which led to a tragic suicide. A whole host of other cases in different areas of science have previously garnered attention in the media, for example the Schön scandal. Overall, this suggested to me that a story exploring scientific misconduct, but told by an insider, a scientist, might be worth telling and might be of interest for non-scientists to read.

© Springer International Publishing Switzerland 2016
P. Rørth, *Raw Data*, DOI 10.1007/978-3-319-23974-3_2

For scientists, misconduct is something we are very aware of, at multiple levels. We are taught how science should be done and we understand the principle, so we know what is correct and what not. To formalize this, and no doubt also to protect themselves, most institutions have in recent years implemented mandatory ethics courses. Scientists know the harsh penalties for misconduct: basically, being banished from science. This is devastating. For an ambitious scientist it is probably far worse than any monetary fine. Scientists obviously read about the spectacular stories like everyone else, but we also hear about plenty of less spectacular ones. We read retraction notices and hear of colleagues getting in trouble one way or the other. We hear of the occasional unpleasant self-righteous zeal in those pursuing such cases. On the other hand, scientists also understand where the temptation comes from and how easy it might be to give in. We know how difficult it can be to detect such events. Finally, we are aware of the tricky gray-zones. Over-hyping a conclusion is usually not actual misconduct. Ignoring a bothersome piece of data might be. Getting a little too inspired by someone else's findings and ideas is just part of the competitive game (and probably why people rarely present unpublished data at meetings anymore). But presenting it as your own is off. And so on.

For the story, I deliberately chose a mild case of misconduct—a one-off fudging of a result in a context where everything else done correctly. The perpetrator, Chloe, is someone who is otherwise an excellent scientist. She is not a psychopath who lacks a sense of reality, just someone who in a moment of weakness chose to do the wrong thing. Given that all scientists know misconduct is wrong and given the harsh penalties for being found out, the question is why an obviously intelligent person would ever do it at all. This is one thing I wanted to explore in the story. Other characters do things that are not exactly laudable, even malicious, so it is not about one character being "bad". There is also a question of degree of responsibility of the head of the lab. The Tom character did not know exactly what Chloe did. But should he have known? Could he have? Perhaps surprisingly, there is no simple answer to that. Scientists working under related circumstances will recognize the dilemmas.

Zooming back out—the consequences of misconduct of this sort on the progress of science are relatively modest. When it is found out, it is corrected by a written retraction. In the modern world, retractions are linked online to the original paper, so readers can't miss it. In the story, the result wasn't even really wrong. This difference between procedural wrong and true results creates some extra wobbles. But science and the scientific approach rest on absolute respect for the data obtained in experiments. Once you fiddle with that, thinking you know what the result "should be", it is no longer science. This is why there is zero tolerance for this type of behavior, even if the scientific consequences are minor—as in the story, where the faked result was essentially correct. High-profile cases can have other, more severe consequences for the scientific community. If the story spins out of control in the media, all of science looks bad. It is hard to insist on the absolute value of the scientific approach, if it is sometimes abused by the scientists themselves. I agree with Stuart, Tom's friend, when he says that most people are honest, so science does work. And the checks and balances of the collective scientific enterprise means mistakes get corrected—eventually. But sometimes it is messy.

2. There is a claim that many—or most—published results in biomedical science are not reproducible and thus worthless. Some even say that peer review of scientific papers does not work and so should be scrapped. What do you think?

Let's take peer review first, specifically as used to evaluate papers for scientific journals. It functions differently in different places, but usually involves a journal editor and a couple of referees. The editor will generally have broad knowledge of a field and can either be a dedicated professional or an active scientist who also runs a lab. Referees are usually experts in the field of a paper, i.e., they are peers of the author. Ideally they are sensible and objective, but they may also be highly opinionated or even be in competition with the author. Direct competition generally has to be disclosed as a potential conflict of interest. With input from the referees, the editor determines whether a paper is appropriate for publication in their journal. Is it factually correct, truly novel and important enough? In addition, referees usually ask for modifications to improve the paper if publication in the journal is possible. These may be minor adjustments or, typically in the field of biomedicine, additional experiments to support the thesis of the paper; in other fields these may be requests for further computational studies to validate theoretical modeling. This feedback can be very constructive and, in my experience as author and editor, often improves papers. But it is not a perfect system. There are lots of challenges for peer review discussed in numerous articles about the subject. These include how best to ensure veracity of data and statements, while making sure busy scientists can find the time to review thoroughly. Also, how best to assign "significance" to new work. My years as Executive Editor of the *EMBO Journal* have taught me just how challenging this task can be. Perception of significance can be quite subjective and often experts simply disagree. But even if imperfect, I think it would be very bad for science if peer review were scrapped. One should consider the alternative: No peer review. Every scientist could publish/post what they want. In areas with many publications, how would the reader choose what to read and what to trust? If every statement of "X shows that Y" is made without having to defend the conclusion to a mini-jury of referees and editor, such statements might become meaningless. A reader who is an expert in the subject can judge, but no one else. Knowing their work has to pass critical peer review, researchers are more careful with what and how they write. Furthermore, if no one checks novelty, publication lists would become endlessly inflated. Online open commenting on papers is not a realistic solution, as far as I can see. I trust that scientific communities will continue to come up with improvements to peer review. The challenges differ for different areas and shift over time, with changes in technology, complexity, habits. But the general principle that scientific conclusions are checked and evaluated by neutral experts and editors, and not just by self-selected scientific friends, makes good sense to me.

The claim of limited reproducibility of biomedical publications is, at face value, shocking. It suggests that most science in this area is simply wrong. In my view, this problem is largely unrelated to issues with peer review, as peer review in this area does not (and really cannot) involve reproducing experiments. Lack of reproducibility gets noticed, for example, when a company decides to repeat every published result that an expensive drug-development program depends upon. I am no expert on this subject. But practically any active scientist builds their work to some extent upon published results from others. This means we find out whether these other results are valid. We also have ample second-hand knowledge from friends and colleagues who rely on yet other sets of data from yet other labs. Based on such anecdotal information, I think most published results are basically correct. But some are not reproducible. Mistakes and, oddly enough, luck both contribute to the problem. The former is obvious. The latter is related to biological variation, the limitations of statistics and the fact that in lab science, you can publish positive results, supporting a conclusion, but it is difficult to publish confounding or negative results. So if you get the latter, you drop the idea and change direction. This biases what is published. But one should also bear in mind that the conditions may be subtly different when re-testing is done. Subtle differences in conditions can have major effects on outcome. Anyone who has moved their lab knows this—things that worked routinely before suddenly don't. Another example is patient trials. It is not unusual for a new study to have different conclusions from a previous one, even if both are statistically solid. Under lab conditions, things should be more reproducible, but they are not perfect. Some areas of lab science are particularly sensitive to subtle effects, like behavioral effects of genetic mutations. But even simple survival of identical mice under controlled conditions shows variability. This is the bane of biology. Sometimes different outcomes can be traced back to experimental differences (minor difference in genetic background or physical environment). In other cases, they remain unexplained. Overall, I think unaccounted-for experimental differences and selective publication of what works, coupled with over-generalization, are primarily responsible for the observed lack of reproducibility. It can also reflect fraud—misconduct—but I think this is rarer. It is irritating and wastes time. But mistakes are eventually weeded out from the scientific literature, whereas true findings form the basis for other fruitful experiments and thus eventually become reinforced as part of our core knowledge.

3. Publish or perish is a well-known saying and is valid for science as well as other academic pursuits. In the story, the protagonists are also very concerned with which journal they publish in. Why is this?

For scientists in academia publication means everything. This is how you officially present your findings and stake your claim to these findings. This means timely publication is important. Should someone publish a finding before you, you are "scooped". However, simply publishing an article in a respectable peer-reviewed journal is not what most scientists want, not in the fields I know. Practically everyone wants top papers, in the top journals. This means the journals like *Nature* and *Science* (and *Cell*), as referred to in the story, and the high-end journals the next level down. There are good reasons for this. Getting a paper accepted in a top journal means that the referees and the editor find the work to be very important. So it is a serious stamp of approval. In addition, with the vast number of papers published in some areas, even the most diligent scientist simply cannot look at everything. But everyone sees what is presented in the top journals. So consider the young scientists like Karen and Chloe in the story. They need for their work to be noticed, so they can stand out amongst hundreds of applicants for principal investigator and faculty positions. One top paper can make a huge difference. Perhaps not surprisingly, focus on top journals can therefore become quite extreme. People will sometimes delay publication of results for years in order to develop a more complete story that has a chance at a top journal. They then have to satisfy the referees' demands. They may try several journals, one after the other, to see who can be convinced. This takes time and effort—and you might get scooped in the meantime. So it is a risk. Risk-taking can be OK if, as is the case in this story, the person with the most to lose and gain, the first author postdoc, makes the decision about where to publish. But if someone else makes that decision— an ambitious lab head, for example—then the situation can become much more conflicted. If several members of a lab work on one project, there are additional problems, of course. The emphasis on journal names, on branding, has other effects on the science publication landscape. But that is another discussion, I think.

4. You mention getting scooped, someone publishing a story before you. It comes up several times in the story. Does it happen a lot? Is it a problem in science?

It depends on the sub-area. If an area is popular it will be densely populated, and the chance of someone else finding more or less what you find is of course greater. This can be very hard for the person being scooped, especially early in their careers when they don't have other published work to fall back on. What happens to Karen in the story is not so common—being scooped on what you think is a unique finding in a new area by a lab you have never heard of. What happened to Chloe as a PhD student is more typical. She found a new element in a well-studied pathway. As the missing link, it was considered important enough to be publishable in a top journal. But others were working on this pathway as well and someone else found it at the same time. One might think that fear of being scooped might then cause scientists, especially younger ones, to shy away from crowded areas. However, the opposite is generally true. If an area is popular and crowded, more PhD students (and postdocs) want to work on it. This attraction is at least in part derived from the notion that what is "in" must be interesting. Perhaps paradoxically, it can also make sense in terms of publishing. A popular sub-area means many articles being published in a short timeframe, primary papers as well as reviews. This, in turn, means many citations, quickly. Many citations to a paper within 2–3 years of its publication is what journals want, to boost the journals' impact factors. So journals may consciously or unconsciously give that area preferential treatment. This means there is positive reinforcement of an area deemed to be "hot". Competitiveness is of course also a driver and a stimulant in itself, so popular areas will tend to move fast. This gives excitement. People will want to hear about it, so the significant players will be invited to give talks frequently. An unexpected positive aspect of crowdedness is correctness. The "waste" of multiple labs pursuing the same ideas also means conclusions are quickly double-checked. And if mistakes are made in the rush to get published, they won't stand uncorrected for long. Other labs will quickly use any published information deemed significant. And they will correct it—with pleasure, most likely—if it is wrong. So overall, while the individual cases of being scooped may indeed be very painful, the fact of this happening and of having crowded areas is not detrimental to science.

5. What is the role of scientific journals in science? Might the high-end journals (unintentionally) be promoting misconduct? They seem to have many retractions.

The original role of scientific journals was distribution of articles to potential readers. With the Internet, this has changed. Because it is now possible to post and distribute your own articles for all to see, the second role of the journals as quality filters has now become their raison d'être: evaluating and selecting articles as "good quality" and "important", based on peer review. With regard to misconduct, the story presents a fairly typical situation that can nudge someone in the wrong direction: Chloe's paper is close to being accepted at a top journal, a coveted prize. But the referees demand that an additional experiment is carried out. There are also time constraints, so the pressure is on. Since the result that the referees want from this experiment is very clear, the temptation to cut a corner is obvious. But asking for an experiment (and, in a way, for a specific result) does not mean the journal or the referees are promoting misconduct. Maybe it is a bit like leaving the jewelry in the shop window overnight, though. It does not make a thief, but it does make theft more likely. Sometimes, extra experiments requested by referees can be insightful and add value to a paper. But they can also be, or seem to the author to be, arbitrary hurdles that just need to be dealt with. But to answer the question about high-profile journals and retractions directly, I don't think it reflects journals or referees being lax about standards. More likely the extra high pressure to get that one paper accepted may encourage exaggerated statements or cutting of corners. The high visibility of the journals no doubt increases the scrutiny of papers and results as well.

As an aside, I would like to point out that scientific journals only have the power that scientists give them. Evaluation of who gets an academic job in science, of who gets grant funding, awards and so on, is usually done by peer review, by other scientists. If we, as scientists, use the journals' decisions on which papers to publish as direct evidence of someone else's worth, then we choose to empower the journals, that is, the editors and their chosen referees. The pressures I talked about grow correspondingly at top journals. If a field of science is so small and specialized that everyone knows what every else in the field has found, published or posted online, then journals become largely irrelevant. Knowing the right people may be everything, instead. But in large, broad fields like the biomedical area, with so many publications, pre-selection of some sort is needed to identify people and papers of interest. Making use of the peer review already carried out by journals seems sensible. The difficulty is in comparing the value of publications in different journals. Sometimes this becomes simplified to looking at the journals' impact factors. The pitfalls of this is a whole discussion of it's own. Suffice it to say that trendiness tends to get overvalued this way. For big decisions, such as who gets an academic job, presentations and personal interviews also matter a lot. The chapter with Chloe's job interview is meant to give a feel for someone doing well in that challenging process. But, of course, significant papers are usually needed to get the initial attention, to get the invitation for that all-important talk and interview.

6. There have been discussions recently about inappropriate "spin" and hype in presentation of scientific results. It contradicts the idealistic view of objectivity in science. How big a problem is this in, and for, science?

New findings and ideas can easily be overlooked by busy readers if they are not presented well. There are, of course, eminent scientists who come to be respected and acknowledged despite writing papers that are difficult to understand due to presentation style. But I think they are exceptions, especially in present-day busy times. Sometimes good ideas can even end up being partially attributed to someone with a better gift for communication than the originator. So, yes, in science, as elsewhere, effective presentation of ideas and their potential significance is important. But there is a big difference between presenting well and over-selling. There can be a lot of excess spin and hype, and I think we see more of it the closer we get to subjects of popular human interests—curing cancer, explaining sexual preferences etc. Authors, institutions and funding bodies have an interest in drawing attention to their work. Can this lead to overt misrepresentation of the significance of a finding? Yes, but I believe this is less common. More frequently, at least in the cases familiar to me (the "X cures cancer" type), what happens is a gradual bending of the truth. To get a paper published or get to a grant funded in the biomedical area, you have to explain what the work is good for in terms of added understanding or potential disease treatment. Scientists authoring a paper may add a few positive general statements to this effect that are not untrue, but that may be a bit optimistic. Other scientists are used to this and take such statements with a grain of salt. But someone writing a press release may well further over-simplify or over-state the claims. Their job is to get the press release noticed and taken up by the press; they do not get points for subtlety. The journalists then build on this to make catchy headlines. It finally ends up being quite distorted. For example, I have been reading about one "breakthrough" after the other in cancer research and treatment on the BBC website that I can only shake my head at. They are not directly untrue, just really misleading. It is frustrating for me as a scientist to read.

That being said, most science is gratifyingly real and solid. There are plenty of real discoveries that do make a real difference. This has always been reassuring to me—and I would imagine many other scientists as well: there is spin, yes, but also substance. Spin, exaggeration, simplification is probably unavoidable as part of human nature. But as scientists, we are lucky to be working in an area that at least also has substance. Many—perhaps most—human endeavors have a much higher ratio of spin to substance.

7. Personal ambition plays a strong role for all three protagonists in your story: Karen, Chloe and Tom. Envy also rears its ugly head. Is science driven more by ambition than by natural curiosity?

Most successful scientists have both drivers, I think. On one hand interest and passion for their science and on the other, straightforward ambition. Scientists generally are a bunch of curious and analytical nerds. Many of us were as children curious about the natural world around us and were stimulated by reading and learning about it later on. The curiosity can remain very strong throughout a career. There are always new questions to ask, another level of understanding. Curiosity gets a lot of people into science. But then there is making a successful career of it, that takes additional skills, generally.

Personal ambition can be displayed in obvious, even unpleasant, ways. It can be more nuanced, or even hidden, in more personable types. But I am convinced it is a very significant forward-driving force in our society. The world of science is simply no exception. My protagonists are people who all "have what it takes" to become successful scientists, including ambition. Ambition matters in part because it drives you to work hard and do your best, day after day. The scenario I have set up, where accomplished postdocs all run their own projects in a high-power lab, is quite realistic. So if these postdocs are not driven, much less will happen on the projects and they are unlikely to succeed. They need to be able to push themselves past the tough periods and the setbacks. The other postdocs in Tom's lab, those we only get a glimpse of, are also ambitious. Different sorts, different quirks, but they have that in common. These young scientists also push each other, stimulate each other. Tom contributes and nudges people along as well, but the postdocs are the drivers. There are other ways labs can be run. Young scientists may more directly be told what to do, pushed ahead, by a forceful lab-head. If the young scientists in a lab are not self-driven, it may be the only way a lab can be productive. The lab I portray in the story represents a subtype that I think many scientists will recognize.

The less pleasant bedfellows of ambition, excessive competitiveness, envy, even destructive tendencies are, quite naturally, present in the story as well. The push of having plenty of accomplished, smart and ambitious people around every day will stimulate these negative features as well, especially on a bad day. Insecurities and self-doubt are part of the make-up of driven people, I think. It comes out in different ways for Karen, Chloe and Tom, but it is there. Scientists are just people.

8. The nicest characters in the story are no longer active in science. They have jobs as teachers or journalists. You even use the term "failed scientist". Why?

It is true that the ambitious protagonists are not the nicest of the characters. It makes sense, I think. Some of the qualities that make people successful in science or any other competitive profession are not what one would call nice. Bill, Karen's husband is nice: gentle, understanding and mostly supportive, although he does not like Karen's competitiveness and obsessions. He is happier being out of that world, and is now teaching. Whether he is more or less smart than Karen is not the point—we don't know. But she is the one who can and will fight her way to a career in academic science, despite her self-doubts. The journalist, Frank is a minor character, but also seems at ease with himself. It is obvious that good teachers and good science journalists are incredibly important for our society. But academic scientists can be terribly snobbish about who is important and who is not. The "failed scientist" term is awful, I know, but I included it on purpose. I have heard this phrase used quite frequently, in particular from irritated authors making condescending remarks about professional science editors. I happen to have worked very closely with a group of such editors the five years when I was in charge of the *EMBO Journal*. I respect them and their work tremendously. They have opted away from the straight academic path, but are still deeply involved in science, thinking about scientific insights, all the time. Some research-active, academic scientists are brilliant and admirable. Others are not. Some science is sparkling and inspired, some is not. Derivative, unoriginal work can be very worthwhile if it has an important practical application. Value comes in different forms. The important point is that being an active, paper-producing scientist, is neither better nor nobler than anything else. This is why I deplore the "failed scientist" remark.

9. The head of the lab, Tom, does not deal very firmly with the most destructive character, Christopher Turrell, now or in the past. He also chooses not to retract the retraction, despite new data. Why? Are these correct choices?

Although Tom is portrayed as a good lab head, he is as much driven by his own interests as the younger protagonists. With respect to Christopher Turrell, he originally did more or less what he could within the norms of that environment. Disruptive, mean-spirited but smart people are generally not easy to deal with. Tom would also prefer not to think too much about his own shortcomings with respect to Dan and his project. Christopher reminded him of that—another reason to avoid dealing with him. It would be exceedingly difficult to do anything effective against Christopher's malicious online article, in particular as it is partially based in reality via Chloe. So Tom does not fight Sushma's cool assessment of "leave it alone". His primary concern is to make the negative publicity go away as quickly as possible. And by not doing much, he succeeds quite well. When there is a chance of serious damage to the lab, a full-scale investigation in case the ethics committee decides to follow up on Chloe's case combined with Christopher's complaint, Tom does do something. He does the smart thing. He gets Chloe to confess. She is ready, as it happens, and it is ideal for Tom. Later on, when Karen comes with new data regarding Chloe's paper, he does not want to draw further attention to the story. You cannot really retract a retraction, but they could have made a 'smaller' publication with new data. Objectively, it would be better for the scientific community to know these new facts. Instead, out of self-interest, Tom pushes this away as well. Combined with positive interest in Karen's project, he manages to steer the situation so no one ends up being too upset. It is not by accident that Tom is a successful person. He is a quite savvy operator.

10. Are there other themes of the novel that you would like to emphasize?

Foremost—how amazing and inspiring doing science can be. It is not just tough and competitive, but also wonderful. I do hope that comes across to the reader. The excitement, the thrill, is in the little day-to-day things, like seeing something new and unexpected and potentially important under the microscope. Karen and Chloe describe these pleasures and thrills in different ways when they talk to Ashok and Frank, respectively. It is also in the bigger things, of course, beyond the lab: Seeing something you have done changing how a particular aspect of science is understood or done. Seeing your work be appreciated by scientists you admire. It is good for the ego, but also makes you believe you have contributed something of value. I hope that this excitement feels real to the reader. That was one of my main motivations for writing the novel: I wanted to give non-scientists a chance see what life in science is like, how captivating it can be. Scandals may make headlines—and story-plots—but they are not the essence of science. The real thing is the day-to-day pleasure of figuring out a good experiment, seeing a nice result, seeing it reproduced, getting a new idea after an old one loses luster—I could go on and on. . . . I feel very lucky, and privileged, to have had an interesting career in science. I have worked in stimulating places and have met some fascinating, impressive people. It's been fun.

11. What parts of the story are based directly on your own experiences?

A lot of the everyday things that go on in the lab are based loosely on my own experiences and those of people I know, but only very loosely. The individual characters and the specific things that happen are all fictional. Even the science stories, and by that I mean Chloe's and Karen's findings presented along the way as well as the published papers referred to, are fictional. They are close enough to real science to be realistic, but that is all.

There is one personal experience that I used deliberately, but in a modified form. A few years ago, I had to retract a paper. There was no misconduct in the person's work or any hint of bad intent, just an unfortunate set of mistakes that led to the wrong conclusion. As mentioned previously, retraction of a paper can be due to misconduct, but it can also be due to straightforward mistakes. In the latter case, the purpose of a retraction is to set the record straight and to inform the scientific community. Well, because of the mistakes, the main conclusion of our paper was wrong and it had to be retracted. Not long after, a "magazine" ran a commentary on the retraction. The article was not malicious, but it was in a place that gets more readers than most scientific articles do. So pretty soon, an entry with the word "retracted" was very high up on the list of

Google search hits when searching with my name. That was a very unpleasant surprise. If you clicked on it and read it, you would not find any of the nasty business I have in the novel. But it was highly visible. Having experienced this, I can better understand why people try to avoid corrections and, in particular, retractions, even if they sometimes should make them. Who wants the unwelcome attention?

12. Finally, we often hear about under-representation of women at the PI and Professor level in science, and possible discrimination. Having been active in science for many years, and being female, what are your thoughts on this?

In biomedical sciences, in most or all countries, the fraction of female professors is much lower than the fraction of female postdocs and PhD students. In many places, the latter is more than half, which is different from physical sciences and engineering. These are facts. The key question is whether the attrition at senior levels reflects a widespread, imposed bias against women succeeding. In other words, is this a problem that needs to be addressed by direct, "affirmative" action? I don't think so, not in the present-day world of science that I know. I should probably point out that my view on this issue is not very popular.

Female under-representation at the top is partly historical: today's full professors were students many years ago when ratios were different. That is easy to correct for. In my opinion, another large factor is personal choice. As discussed previously, success in science requires ambition and drive as well as ability—choosing to give it your all in a competitive world. Men seem more likely to consistently make the straight, career-optimizing choices. Not all, of course, I'm just talking averages here. I took a competitive route and did well. I have now made a choice to leave science—this "hurts" the full professor gender statistics. Others make different choices along the way for various reasons—I'll get back to that.

For about 25 years, I have worked as a scientist, all over the world. I have never felt discriminated against or disadvantaged by being female. What I have felt in recent years, however, is a pervasive pressure to consider gender when making decisions in science, instead of just considering the quality of people's work. In planning of meetings, in committees, in grant reviews, in scientific evaluations of institutes, the question of whether there are enough women comes up again and again. If not, then we have to correct it. But if 20 % of the independent investigators in a field are women, then requiring 50 % of speakers at a meeting to be women is neither fair nor good. While well-intentioned, forcing a "good" gender-ratio subtly brings along the message that there are two classes of scientists, to whom different sets of evaluation criteria apply. This is not a good thing. As a scientist who happens to be born with two X chromosomes, I feel diminished by preferential treatment. It is condescending. In a professional context, I want to be treated based on merit, not gender. I want to be allowed to do the same for others. I doubt that I'm the only person who feels that way.

I'm not saying there is never any bias. There have been studies showing examples of negative bias against women. But I also know of a large, well-designed study showing exactly the opposite. This study was never published. Instead, for whatever reason, it was allowed to be quietly forgotten. Political correctness is powerful; things can become very unpleasant for anyone who dares to stand up against it.

A related and important issue is that of balancing time between a demanding career and family, especially with young children. This is difficult in science, as it is in any other competitive career. I don't have children myself, but I do have nieces and nephews, so I get it. There are only 24 hours in a day and all people, male or female, have finite mental energy. This gets us back to personal choices. But it is a social issue and should be discussed as such. For example, it is completely irrelevant for women (or men) without children. For professional evaluations, less focus on people's age and rapid career progression and more on quality of recent work might help. Getting rid of tenure is a possibility. It seems ridiculous to me that academic employment from years 40 to 65 should depend almost exclusively on work done from years 25 to 40. Society and institutions can also contribute by ensuring access to affordable help with practical stuff. For high-pressure, dual-career marriages to function with kids, help is generally needed, such as full time nannies. Paradoxically, this is easier to get in countries with an uneven social structure and low minimum wage, like US or Singapore, than in socially more fair countries, like Denmark. Men could do more, if they want to have children. Ambitious women who want children could try harder to be attracted to less ambitious, potential stay-at-home dads. It's a division of labor and responsibility that appears to work well for many ambitious men. There are lots of angles to consider. But the point is, this is not an issue of negative professional gender bias and it cannot, and should not, be "fixed" by preferentially selecting female candidates for jobs or other professional functions.

Anyway—enough said about that, I think. These are just my opinions, based on my own experiences in science. Others may disagree.

Printed in the United States
By Bookmasters